# When God Was a Bird

groundworks|

ECOLOGICAL ISSUES IN PHILOSOPHY AND THEOLOGY

**Forrest Clingerman and Brian Treanor,** *series editors*

# When God Was a Bird

## Christianity, Animism, and the Re-Enchantment of the World

*Mark I. Wallace*

FORDHAM UNIVERSITY PRESS

*New York 2019*

The illustrations found in this book are by James Larson and reproduced with his permission.
Fordham University Press has no responsibility for the persistence or accuracy of URLs for external or third-party Internet websites referred to in this publication and does not guarantee that any content on such websites is, or will remain, accurate or appropriate.

Fordham University Press also publishes its books in a variety of electronic formats. Some content that appears in print may not be available in electronic books.
Visit us online at www.fordhampress.com.

Library of Congress Cataloging-in-Publication Data available online at https://catalog.loc.gov.

Printed in the United States of America

21 20 19 5 4 3 2 1

First edition

*to Audrey*

CONTENTS

# PREFACE

At one time, God was a bird. In ancient Egypt, Thoth was the long-billed, ibis-headed God of magic and wisdom. Winged divinities populated the pantheon of Greek antiquity: the half-bird-half-beast griffin built its nest in mountain redoubts to protect itself from bandits, while the phoenix was a firebird that, upon death, could reanimate itself though the ashes of its own burned body. According to old Norse legends, the principal father-god Odin was also the raven-god who could fly across the oceans and into the underworld with his two black-bird companions. In the theological hierarchy of the pre-Columbian Aztecs, Quetzalcoatl was the plumed serpent deity who cocreated the cosmos. Even today, Quetzalcoatl is still revered in the form of the resplendent quetzal—arguably the most beautiful bird of the Americas—whose green, blue, and red feathers perfectly match the spectacular plumage of its namesake divinity. In our time as well, the Yazidis, a long-suffering religious minority in Northern Iraq, pay obeisance to the Peacock Angel, the Supreme God's primary emanation in the form of the peacock, which, like the resplendent quetzal, is the cosmogonic origin of all creation.

But in spite of—or better, to spite—this time-honored wealth of divine avifauna, the religions of the book, Judaism, Christianity, and Islam, divorced God from the avian world in order to defend a pure form of monotheism uncorrupted by beaks and feathers. In the case of Christianity, biblical faith became synonymous with the destruction of the pre-Christian gods identified with all manner of animals. According to legend, when Saint Patrick went to Ireland, he became outraged at the cult of serpent worship among the Druid priests—and thereby drove the entirety of the island's snake population into the Celtic Sea with his mighty staff. Similarly, Spanish missionaries inveighed against Native American reverence of the California condor as an airborne divinity in an effort to purge new converts of any traces of belief in God as a winged being.

Recently, I journeyed to see the Virgin of Montserrat (Catalan for "serrated mountain") in Northern Spain. With my wife, Audrey, and our friends Steven and Adrienne, we wound our way up the jagged cliffs by funicular railcar to the historic pilgrimage site. At the mountain's summit, we walked to the back of the monastery church and entered a high-ceilinged room limned with gold and precious stone. There we paid respect to an old statue of Mary blackened with age and candle smoke, holding Jesus in her lap—and sitting on an ornately carved chair with a magnificent iron-cast winged creature flying just over her head. In Christian iconography, this feathered being is the Holy Spirit. As the third member of the Godhead, the aerial Holy Spirit, along with the Father and the Son, constitutes the Trinity. Is it significant that from the time of the Bible to the present the central depiction of the Spirit of God is as a bird, specifically, a dove? Throughout the history of Christianity, the Holy Spirit, as in Montserrat, has soared through the air in churches worldwide—in gleaming plumage and airy flesh—in carved objects, wall paintings, colorful icons, pulpit ornaments, altarpieces, rood screens, roof bosses, and illuminated manuscripts. It is easy to dismiss such items as merely decorative. But I believe that these representations of God on the wing—in church art as well as the biblical texts—are traces of the historic Christian belief in God as a bird, in spite of conventional dismissals of such beliefs as not conforming with accepted dogma. As an exercise in theology, philosophy, nature writing, and personal anecdote, this book is a recovery of the lost veneration of the bird-God within Christian thought and culture. My hope is not only that this recovery will *return* biblical religion to its earthly origins but, moreover, that it will *enflame* the desire to love the Earth as God's earthly nest—indeed, as the cherished site of God's, and all beings', daily habitation.

*Mark I. Wallace*
*Crum Creek Watershed*
*Delaware River Bioregion*
*In honor of the Nanticoke Lenni-Lenape Homeland*
*Easter 2018*

# Introduction

## *Crossing the Species Divide*

### *The Animal God*

In a time of rapid climate change and species extinction, what role have the world's religions played in ameliorating, or causing, the crisis we now face? It appears that religion in general, and Christianity in particular, bears a disproportionate burden for creating humankind's exploitative attitudes toward nature through otherworldly theologies that divorce human beings and their spiritual yearnings from their natural origins. In this regard, Christianity today is viewed as an unearthly religion with little to say about everyday life in the natural world. Because it has focused on the salvation of human souls, it has lost touch with the role the verdant world of animals and plants, land and water, plays in human well-being. In principle, Christian belief in the incarnation of God in the human Jesus renders biblical faith a fleshy, this-worldly belief system. In reality, however, Christianity is still best known for its war against the flesh by denigrating bodily impulses as a source of temptation and by dismissing the material world, while not fully corrupted, as contaminated by sin and inimical to humans' destiny in a far-removed heaven of bodiless bliss. As

Sean J. McGrath puts it, in traditional Christian thought, "matter was no doubt good, but not that good, and in its tempting quality it posed a grave threat to the soul: best to have as little to do with it as possible."[1] My book argues that this picture of Christianity as hostile to the creaturely world, while accurate to a point, misses the supreme value biblical religion assigns to all of the denizens of God's good creation, human and more-than-human alike.

Moreover, I argue that this picture, in particular, misses the startling portrayals of God as the beaked and feathered Holy Spirit, the third member of the Trinity who, alongside the Father and Son, is the "animal God" of historical Christian witness.[2] Appearing in the Christian scriptures as a winged creature at the time of Jesus' baptism, the bird-God of the New Testament signals the deep grounding of archi-original biblical faith in the natural world. But due to the age-old bias in world-denying Christianity that God is divorced from mortal existence, this reality of God in creaturely manifestation—not only in the mode of the human Jesus but also in the form of the birdy Spirit—has been missed by most Christian thinkers and practitioners alike. This lost truth is a hidden pearl of great price. In *When God Was a Bird*, my aim is to correct this oversight in contemporary religious thought and pave the way for a new Earth-loving spirituality grounded in the ancient image of God as an avian life-form.

In the history of Christian thought, Trinitarian portrayals of the Spirit eloquently make this point: the Father and Son are depicted in human familial terms, while the Spirit is figured as the avian divinity who mediates the relationship of the other two members of the Godhead. My recovery of God's animal body within biblical and Christian sources might be startling, even sacrilegious, for some readers at first. Even though the Bible speaks directly about God as Spirit becoming a winged creature ("When Jesus was baptized, the Holy Spirit descended upon him in bodily form as a dove"; Luke 3:21–22), religion and biblical scholars alike have oftentimes dismissed the descriptions of God's Spirit as a bird in the New Testament as a passing allusion or figure of speech. These critics do not regard this and similar texts as actual descriptions of the avifauna that God became and is becoming: testimonies to the Spirit's enfleshment (or, better, enfeatherment) at the time of Jesus' baptism. Nevertheless, I maintain here that the full realization of Christianity's historical self-definition as a scriptural, incarnational, and Trinitarian belief system is *animotheism*[3]—the belief that all beings, including nonhuman animals, are imbued with divine presence. Buried deep within the subterranean strata of the Christian witness is a trove of vibrant bodily images for God in animal form (as well as in human and plant forms), including, and especially, the image of the avian body of the Holy Spirit. Woven into the core grammar of Christian faith, then, is the belief in the Spirit as the *animal* face of God, even as Jesus is the *human* face of God.

Though I will note in Chapter 1 one expression of the vegetal embodiment of God in the Exodus story of the burning bush, my overall focus in this book is on the premier animal form of divinity in the Bible and Christian thought, namely, the Holy Spirit, the bird-God of classical evangelical witness. But to focus on the Spirit as God's animal modality is not to deny God's many botanical incarnations within Christian scriptures and traditions. For the botanist Matthew Hall in *Plants as Persons: A Philosophical Botany*, Western religion is unreflexively "zoocentric" because it appears only to value certain sentient beings (human and non-human animals) over and against the numerous plants that populate our daily lives.[4] But there are many scriptural counterpoints to Hall's broadside. To take one, consider Jesus' paean to the spectacular wildflowers that graced his pathways in biblical Israel/Palestine: "Consider the lilies of the field, how they neither toil nor spin, and yet I say that Solomon in all of his glory was not clothed as beautifully as these lilies" (Matthew 6:28–29). Modern biblical scholarship assumes "Solomon" in this instance is a metonym for the lavish tenth-century BCE Jerusalem Temple and Palace attributed to King Solomon. If this is the case, Jesus' analogy is stunning: the beauty of commonplace lilies is a more fitting expression of God's earthly habitation than the actual built tabernacle that housed Yahweh's presence in biblical Israel. As the site of divinity, Jesus' green religion valorized the vegetable world as much as the animal world. Hall's posthumanist analysis of some forms of Earth-hostile religion is much needed. But his overall critique of biblical spirituality misses the point. Hidden in the bedrock of Christian theology is a grounding *animist* sensibility that construes all things—including the sentient and relational biomass that makes all life possible—as living enfleshments of divinity in the world.

I will call this new but ancient vision of the world *Christian animism* in order to signal the continuity of biblical religion with the beliefs of Indigenous and non-Western communities that God or Spirit enfleshes itself within everything that grows, walks, flies, and swims in and over the great gift of creation.[5] I hope to revitalize Christian theology with a blood transfusion from within its own body of beliefs and also from global religious communities whose members encounter divinity in all things.

I suggest that this blood transfusion is a genetic match with the deep cellular structure of Christianity because it is a product of that structure itself—as well as being borrowed from other compatible religious traditions. Is it possible to restore Christianity's primordial experience of the world as the enfleshment of sacred power? Can God be seen as ensouling every life-form with deific presence, rendering all things consecrated family members of interrelated ecosystems? This Janus-faced effort recovers the once-lost and now-found essence of the Christian religion. So my question is, is my *ad fontes* effort consistent with Christianity's historical self-understanding, even though the religion today has largely forgotten its primordial beginnings and thereby its originary vision of the world as sacred place, as holy ground, as the body of God?

In Christianity's practiced forgetfulness of its earthbound origins, it has recast itself as a footnote to Greek philosophy. As a vassal to Plato and Aristotle, it has operated within a graded hierarchy of Being in which plants and animals, rocks and rivers, are denigrated as soulless matter, while human beings are elevated as godlike, intelligent creatures—mired in the muck of corporeal existence, to be sure, but still able to shake off the mortal coil that binds them to the lower life-forms and realize their true *imago Dei* natures and destinies. Today, Western Christianity continues to function within this anthropocentric universe and has become a pale and distant echo of its biblical-animist origins. It is for this reason that Christianity has endured, and continues to endure, a centuries-long "Babylonian captivity"[6] to ossified *contemptus mundi* philosophical categories and divisions. This captivity has consistently led Christian thinkers into a Neoplatonic cul-de-sac in which the world is maligned as a dead and fallen place wherein the human soul, divorced from its body, strives to transcend its physical drives and passions and, in so doing, return to the disembodied Source from which it originated. But Christian animism interacts with the world differently—not as a sinkhole of corporeal lust and confusion to be battled against and overcome but as the privileged site of God's daily habitation. *In short, Earth is God's natural home.* Or as the theologian Shawn Sanford Beck puts it, "*Christian* animism, then, is simply what happens when a committed Christian engages the world and each creature as alive, sentient, and related, rather than soul-less and ontologically inferior."[7]

But labeling Christianity as an animist belief system—the conviction that all things, including so-called inanimate objects, are alive with sacred power and worthy of human beings' love and protection—is a misnomer for Christian believers and religious scholars alike who regard biblical religion at odds with, and distinct from, the pagan religions of primordial people. In spite of Christianity's animist origins—or perhaps *to spite* its vernacular beginnings—Christianity viewed itself as a divinely inspired religion of the book that is categorically different from the commonplace forms of religion that showed special regard for sacred animals, tree spirits, revered landscapes, and hallowed seasons of the year. In this telling, Christianity replaced the old gods of pre-Christian animism with the new revealed religion of Jesus, the saints, and the Bible. Correspondingly, it saw itself as a type of pure monotheism vis-à-vis alternative forms of so-called primitive or polytheistic religions that were based in fertility rituals and nature worship. Then and now, Christianity regards itself as an otherworldly faith that superseded heathen superstition insofar as its focus was on an exalted and unseen Deity who is not captive to the vicissitudes of mortal life on Earth.[8]

Challenging the conventional wisdom that Christianity and animism are contradictory traditions, I make reference here to the "religious turn" within contemporary Continental philosophy as a background resource in new studies about God's animal body within biblical sources.[9] As well, the related fields of *posthumanism* (the antispeciest disavowal of human chauvinism) and *new materialism* (the analysis of the agential subjectivity of nonhuman material realities) are also sotto voce dialogue sources in my return to animism.[10] The suggestion that the nonhuman animal is the face of divinity within the plurality of God's many corporeal expressions is characteristic of this religious turn in modern philosophy and related fields of study. The suggestion begins with everyday animals—in particular, cats and dogs—as hints of divine presence in the world. Martin Buber's *I and Thou* sets forth a relational ontology wherein Buber "looks into the eyes of a house cat" and catches the breath of eternal life wafting about him, because in "every You we address the eternal You."[11] Similarly, Emmanuel Levinas asks whether the faces of all others—including all animal others—are intimations of divinity in the world: "One cannot entirely refuse the face of an animal, . . . for example, a dog. . . . But it also has a face. . . . It is as if God spoke through the face."[12] And echoing Buber's encounter with a house cat, Jacques Derrida in *The Animal That Therefore I Am* marks his own vertiginous elision of a cat's discriminating stare and the penetrating gaze of God. Derrida says that standing naked in front of a cat, he hears the cat—that is, he hears God—address him at the core of his personhood: "I often wonder whether this vertigo . . . deep in the eyes of God is not the same as that which takes hold of me when I feel so naked in front of a cat, facing it, and when, meeting its gaze, I hear the cat or

God ask itself, ask *me:* Is he going to call me, is he going to address me?"[13] What ties these philosophical ruminations together is the phenomenon of *being addressed* by other-than-human beings whose yearnings for relatedness is consistent across different orders of being—that is, relatedness among animal others themselves, between animal others and ourselves, and between animal others, ourselves, and the divine Other. These philosophical reflections about the eyes or the faces of animals as mediums of the sacred has informed my attempts to place into conversation Christianity and Indigenous traditions' celebration of signs of the *anima mundi* within all things.[14]

## *Animism*

In philosophy and theology, innovative attempts at forging connections between biblical religion and primordial belief systems marks a sea change away from earlier comparativist studies of "revealed religions" such as Christianity vis-à-vis preliterate religious cultures. Beginning in the mid-twentieth century and continuing into the present, a profound shift has taken place toward a critical understanding of the centrality of animal bodies and subjectivities in the formation of all of the world's religions, including Christianity. This shift moves away from the hoary opposition between pure monotheism and nature- and animal-based religion—an opposition that is bedrock to all of nineteenth- and early-twentieth-century British anthropology of religion, including E. B. Tylor's *Primitive Culture*, William Robertson Smith's *The Religion of the Semites*, and James Frazer's *The Golden Bough* and *The Worship of Nature*. At the heart of this opposition between the modern and the primitive in early Victorian studies of religion, the notion of animism was deployed as a proxy for the benighted epistemologies of first peoples who envisioned the cosmos as an intersubjective communion of living beings, including animal beings, with shared intelligence, personhood, and communication skills. As John Grim writes, "During the late nineteenth century colonial period interpretive studies described communication with animals among indigenous peoples as a failed epistemology. The assumption that only humans know, or a least that only humans report on their knowing, resulted in the long-standing critique of indigenous ways of knowing coded in the term animism. As a means of actually knowing the world, animism was dismissed as simply a delusion, or a projection of a deluded human subjectivity."[15]

Sharing resonances with the Latin word *animus*, which means "soul" or "spirit," the idea of animism was significantly advanced in the modern

West by Tylor's analysis of how first people attributed "life" or "soul" or "spirit" to all things, animate and inanimate. In *Primitive Culture*, Tylor writes, quoting another theorist, that in animism "every land, mountain, rock, river, brook, spring, tree, or whatsoever it may be, has a spirit for an inhabitant; the spirits of the trees and stones, of the lakes and brooks, hear with pleasure . . . man's pious prayers and accepts his offerings."[16] Tylor's study of animism emerged out of an evolutionary, Occidental mind-set that described, at least for Victorian readers, the unusual panspiritist beliefs and practices of first peoples—the ancient sensibility that all things are bearers of spirit. Operating from a settler-colonialist mind-set, Tylor denigrated animism as the superstitious worldview of childlike tribes whose beliefs eventually gave way, in his thinking, to the march of reason and science in "civilized" societies. He writes, "Animism characterizes tribes very low in the scale of humanity, and thence ascends, deeply modified in its transmission, but from first to last preserving an unbroken continuity, into the midst of high modern culture."[17] For Tylor, while animism was characteristic of "low" precivilized cultures, its influence slowly weakened over time as "high" cultures became more literate and scientific.

While the term is tainted by Tylor's colonial elitism (animism is characteristic of "low humanity" rather than "high culture"), the concept of animism is being recovered today based on its analytical capacity to illuminate how traditional people, then and now, envision nonhuman nature as "ensouled" or "inspirited" with living, sacred power. An excellent example of this rethinking is the analysis of the sacred personhood of trees in ancient and contemporary India by the Hinduism scholar David L. Haberman. In *People Trees: Worship of Trees in Northern India*, Haberman redeploys the idea of animism in order to efface the hierarchical boundary lines between human and nonhuman and thereby to position South Asian tree worship as a meaningful exercise in cultivating a holistic relationship with the nonhuman world. For Haberman, Tylor and his ilk's dismissal of animism as childish superstition has bequeathed to modernity the debilitating idea "that we now live in a dead world that is truly animated only by human beings."[18] But Haberman notes that many contemporary social scientists are undermining this in/animate binary by reversing the relegation of animism to primitive ignorance and the elevation of materialism as the agreed-on worldview of enlightened, Western societies. By assigning humanlike capacities to other-than-human life-forms, the natural world now becomes a living field of complicated relationships rather than a dead world of lifeless objects. For Haberman et al., animism trumps empiricism as a superior way of knowing and experiencing the totality of existence. Haberman writes,

> As these anthropologists demonstrate, any earnest consideration of the personhood and consciousness of nonhuman beings leads to a reconsideration of animism, once rejected as illusory primitivism. Without the judgmental and cultural evolutionary perspective of Tylor, which disparages (embodied) animism with the pejorative label "primitive" . . . we find many cultures that treat natural phenomena as "proper persons," [and] the sharp divide between human and nonhuman beings cannot be taken for granted. It also cannot be assumed as universal; other possibilities clearly exist. Nor can it be regarded as part of superior civilized culture, unless we wish to maintain the colonial cultural evolutionary perspective of Tylor.[19]

Arguably, no contemporary thinker has done more to rehabilitate the nomenclature of animism than the comparative religions scholar Graham Harvey. Like Haberman's recovery of animism in the South Asian context, Harvey writes that animism "is typically applied to religions that engage with a wide community of living beings with whom humans share this world or particular locations within it. It might be summed up by the phrase 'all that exists lives' and, sometimes, the additional understanding that 'all that lives is holy.' As such the term *animism* is sometimes applied to particular indigenous religions in comparison to Christianity or Islam, for example."[20] In Harvey's formulation of animism, nature is never dull and inert but inherently alive with the infusion of Spirit or spirits into all things. Here there is no distinction between living and nonliving, between animate and inanimate. Harvey's use of the phrase "all that exists lives" means that nature is not brute matter but always full of life and animated by its movement, weight, color, voice, light, texture—as well as its relational powers and spiritual presence. Nature's capacity for *relatedness*, its proclivity to encounter us, as we encounter it, in constantly new and ever-changing patterns of self-maintenance and skillful comportment, is the ground tone of its vibrant and buoyant energy. As the philosopher David Abram similarly argues, nature or matter is not a dead and lesser thing that stands in a lower relationship to animate spirit but a self-organizing field of living, dynamic relationships: "Yet as soon as we question the assumed distinction between spirit and matter, then this neatly ordered hierarchy begins to tremble and disintegrate. If we allow that matter is *not* inert, but is rather animate (or self-organizing) from the get-go, then the hierarchy collapses, and we are left with a diversely differentiated field of animate beings, each of which has its own gifts relative to the others. And we find ourselves not above, but in the very midst of this living field, our own sentience part and parcel of the sensuous landscape."[21]

Abram and others analyze how Indigenous peoples celebrated, and continue to celebrate, relations with other-than-human communities of beings that are alive with spirit, emotion, desire, and personhood. This ascription of personhood to all things locates human beings in a wider fraternity of relationships that includes "bear persons" and "rock persons" along with "human persons."[22] At first glance, this is an odd way to think, since Western ontologies generally divide the world between human persons, other animals, and plants as *living things*, on the one hand, and entities such as earthen landscapes, bodies of water, and the airy atmosphere as *nonsentient elements*, on the other. But the Native American religions scholar George "Tink" Tinker argues that even "rocks talk and have what we must call consciousness," and then continues, "The Western world, long rooted in the evidential objectivity of science, distinguishes at least popularly between things that are alive and things that are inert, between the animate and the inanimate. Among those things that are alive, in turn, there is a consistent distinction between plants and animals and between human consciousness and the rest of existence in the world. To the contrary, American Indian peoples understand that all life forms not only have consciousness, but also have qualities that are either poorly developed or entirely lacking in humans."[23] Glossing scholars such as Harvey, Abram, and Tinker, I am suggesting that animism flattens commonplace ontological distinctions between living/nonliving or animate/inert along a continuum of multiple intelligences: now everything that *is* is alive with personhood and relationality, even sentience, according to its own capacities for being in relationship with others. As Harvey says, "Animists are people who recognize that the world is full of persons, only some of whom are human, and that life is lived in relationship with others."[24] *All* things are persons, only *some* of whom are human, because *all* beings are part of a community of relationships, only some of whom are recognizable as *living* beings by us.

In general, however, most scholars of religion regard animism as far removed from Christianity, both culturally and theologically. In Graham Harvey's definition of animism, recall his assumption that monotheistic traditions such as Christianity are categorically distinct from animism: "the term *animism* is sometimes applied to particular indigenous religions *in comparison to Christianity*."[25] Likewise, the comparative religions scholar Bron Taylor writes that in spite of attempts to bring together animism, which he calls "dark green religion," and the major world religions, such as Christianity, these traditions have different origins, share different worldviews, and cannot genuinely cross-pollinate with one another in new paradigms of Christian animism such as mine. He writes, "For the most part, in spite of occasional efforts to hybridize religious traditions, most of the world's major religions have worldviews that are antithetical to and compete with the worldviews and ethics found in dark green religion."[26]

This book, however, will argue the contrary, namely, that while the Christian religion largely evolved into a sky-God tradition forgetful of its animist origins, its carnal identity is paradigmatically set forth in canonical stories about the human embodiment of the historical Jesus, on the one hand, and, provocatively, the animal embodiment of the avian Spirit, on the other. Writing as an ecotheologian—or as the history of religions scholar Thomas Berry referred to himself, a "geologian"[27]—my reading of the biblical texts and Christian history will cut against the received misunderstanding of Christianity as a discarnational religion. Brushing against the grain of biblical faith's pronounced opposition to Earth-based religion, I attempt to return it to its true animist beginnings and future prospects. Far from Christianity supplanting animism as a foreign or corrupting influence, I maintain that the religion of Jesus both sprang and continues to receive its vitality from its dynamic origins in and interactions with the animist center of its founding vision. Animism is not peripheral to Christian identity but is its nurturing home ground, its *axis mundi*.

## *Feral Religion*

I first began to speak of "Christian animism" in 2010 in *Finding God in the Singing River: Christianity, Spirit, Nature*, where I wrote,

> Surprisingly and paradoxically, Christianity, which historically waged war against "heathen" fertility and Goddess cultures, can now recognize itself as the bearer of the very earth-centeredness that it initially inveighed against. That Christianity *is* animism and animism *is* Christianity is an insight that is now possible as a result of a new, healed relationship between biblical religion, on the one hand, and earth religion, on the other. The Spirit and the earth are one, the Sacred and the planet are one, God and nature are one—so begins the new adventure in the return of Christianity to its green future as a continuation of ancient Pagan earth wisdom.[28]

Returning Christianity to its "ancient Pagan earth wisdom" reflected an earlier call for a "*revisionary paganism* as the most viable biblical and theological response to the prospect of present and future environmental collapse."[29] But I found in conversation with readers that linking Christianity with paganism, while historically accurate to a large degree, led to confusion. In 2010, I turned to the idea of animism, and generally stopped using the term *paganism*, as a more precise analytical category for making sense of the elective affinity between the broad-based assumptions of both Indigenous communities and Christian theology—namely, that the natural world is a vibrant community of living beings, including seemingly inanimate formations, all of whom are sacred and deserving of care and protection.[30]

I am aware, however, that the notion of animism is a difficult candidate for retrieval because it was invented as a derogatory proxy for the premodern (read: barbaric) worldviews of primordial people. As we have seen, the category seems to be hopelessly contaminated by colonial-era white-supremacist assumptions about the evolutionary differences between first peoples and latter-day Euro-Americans. Is a loaded term that has functioned as an ethnic slur against nonwhites, under the guise of social science, still recoverable today? But like similar religious studies' technical terms that have productively entered the mainstream notwithstanding their pejorative origins—for example, the adjective *queer*, in spite of its long-standing homophobic connotations, has emerged as the preferred nomenclature for the analysis of nonbinary and same-sex relationships in religious studies[31]—the concept of animism now appears to be an important methodological tool for analyzing the *vitality* and *sacrality* of all life-forms within Earth community.[32]

In addition to the conceptual work the term *animism* performs—its insight into the relational character and common personhood of material existence—the term has two other advantages. On the one hand, it is increasingly being deployed by scholars of Native traditions themselves, effectively repurposing the category as a postcolonial mode of inquiry, at some remove from its racist origins, vis-à-vis the variety of relational ontologies that underlie complementary lifeworlds. In this regard, for example, consider the history and philosophy of Native science work by Gregory Cajete. Cajete offers, in my judgment, a nuanced study of the ambiguity and the promise of the notion of animism. He analyzes the negative connotations of the term and its potential for generating productive insights into the common subjectivity of human and more-than-human communities in relationship with one another. Cajete writes, "The word 'animism' perpetuates a modern prejudice, a disdain, and a projection of inferiority toward the worldview of Indigenous peoples. But if, as the French phenomenologist Merleau-Ponty contends, perception at its most elemental expression in the human body is based on participation with our surroundings, then it can be said that 'animism' is a basic human trait common to both Indigenous and modern sensibilities. Indeed, all humans are animists."[33] Cajete's insight that "all humans are animists" underlies a second advantage to the term being rehabilitated for religious studies today: its counterdiscursive capacity to invert the hierarchical power relations between the notions of "Christianity" and "Indigeneity" that characterize popular thinking along with traditional academic study of religion and culture. As Darryl Wilkinson puts it, "The new animism is therefore widely presented as a turn to an indigenous (and particularly hunter-gatherer derived) sensibility vis-à-vis the world, and a potentially corrective model for the West to follow."[34] The model of reality as an animate communion of sacred beings is emerging as a paradigm, distinctly characteristic of originary people, that supersedes the experience of alienation and isolation characteristic of the modern West and the Western Christian imaginary as well. It is this "man bites dog" reversal of epistemological priority—it is the truly global religions of Native societies that perceive the common relatedness of all beings, not the religions of the book—that effectively positions first peoples' spiritualities as archi-original and better able to articulate the intersubjective nature of reality as opposed to historical Western Christianity's dependence on a dualistic, animate/inanimate worldview.

The Chickasaw scholar Linda Hogan writes similarly about the return of animism in Western curricula,

> The introduction of the studies of animism to academe was a surprise to me. I left university to work for my own tribal nation, for the people and for the land. Since then, classes in Paganism and animism have been offered in universities. Hearing this for the first time at a conference, I was horrified. We were killed in great numbers for being called Pagans and animists. Now one of the very institutions that disavowed our original relationships with the environment has studies in its return. Those of us who suffered from the colonizing forces in our lives, and from "cognitive imperialism," are now no longer the ostracized. What once victimized us is now a special area of religious studies. And yet to know that any small part of our knowledge is being taught in colleges and universities is significant, even if it is only a small portion of the intellectual knowledge of our traditionalists. It is, in some way, the fulfillment of the circle of life, as painful as it may feel to many of us.[35]

Like Cajete, Hogan's appreciation and suspicion toward the reintroduction of animism in university settings is significant. But in spite of the gradual recognition of the importance of Native understandings of intersubjective ontologies within the academy, many people of Christian faith today struggle to come to terms with the claim that all things are bearers of sacred personhood. In particular, Christians are uncomfortable speaking about the *animality* of God, even though the belief in the *humanity* of God is basic to everyday Christian discourse. In part, this discomfort likely stems from believers' tacitly ordered separation between humans as intelligent and nonanimalistic, on the one hand, and animals as instinct driven and subhuman, on the other. But it may also be, as some scholars suggest, that animals are held in poor esteem, and certainly not elevated to the prestige of divinity, because they are accorded very low value, if even mentioned at all, in the founding canonical texts of the Christian tradition. Is it the case, then, that the world of animals is relatively insignificant in the Bible? Many contemporary scholars in the emerging field of religion and animal studies seem to think so.

As Laura Hobgood-Oster puts it in her otherwise luminous analysis of animals and Christianity in *Holy Dogs and Asses: Animals in the Christian Tradition*, "Although animals are not prominent in either the canonical or the extracanonical gospels, powerful stories emerge from the relatively unknown extracanonical traditions."[36] Barbara Allen in *Animals in Religion: Devotion, Symbol, and Ritual*, her excellent and comprehensive study of religion and animals, comments similarly on the relative paucity, in her judgment, of animal stories in the New Testament. In concert with Hobgood-Oster, she also writes that while there is considerable animal material in books that were left out of the canon, the Christian scriptures themselves make only minimal reference to animals: "In Scripture animals are present at the birth of Jesus. Within the canon, their role is at times small, but in extra-canonical texts their presence is greater, especially at the Nativity and during the early years of Jesus."[37] Allen concedes that the "Holy Spirit, one of the persons of the Trinity, is represented by a dove,"[38] but it is clear from the context of this reference that she is referring to symbolic and pictorial representations of the Spirit—not that the Spirit, as I have suggested here, is a winged animal and, in that sense, that God in Godself is a dove. Like Hobgood-Oster, Allen suffers from a certain blind spot regarding the thoroughgoing descriptions of God-as-avian-Spirit in the New Testament.

### *God of Beak and Feathers*

It is odd to me that animal theologians such as Hobgood-Oster and Allen appear not to recognize that each of the Gospels and, by extension, all Christian traditions testify to the same reality: at the inauguration of Jesus' public ministry during his baptism by John, the Father spoke and the Spirit enfleshed itself as a dove, forever enshrining in Christian belief the Trinity of the cosmic Father, the human Son, and the animal Spirit. So my book's thesis: Christianity, at its core, is a carnal-minded, fleshly, earthy, animalistic system of belief just insofar as its understanding of the human Jesus (*christology*) and the avian Spirit (*pneumatology*) is rooted in its divinization of human and nonhuman creatures (*animality*). In this telling of the Christian story as *animocentric*,[39] God *is* an animal, without denying the difference between God and animals, because the primary Trinitarian grammar of biblical religion centers on the double enfleshment of God in human and avian modes of being, the Son and the Spirit, respectively.

My suggestion is that divine incarnation is not limited to the person of Jesus but includes the person of the Spirit as well, what I call *double incarnation* or what one might call *libertine* or *promiscuous incarnation* insofar as God in Jesus and the Spirit embraces the fleshly reality of all interrelated organisms.[40] Initially, this perspective that the whole expanse of creation is suffused with divinity seems also to be the case in David L. Clough's thoughtfully detailed and insightful Christian animal theology *On Animals*. Affirming the Johannine maxim that the Word became flesh, Clough writes, "The doctrine of incarnation does not therefore establish a theological boundary between humans and other animals; instead, it is best understood as God stepping over the boundary between creator and creation and taking on creatureliness."[41] But does God's assumption or adoption of creatureliness extend as far as God becoming bird flesh in the dovey Spirit at the time of Jesus' baptism? Apparently, Clough thinks not, by agreeing with an interpretation of Augustine that says, "Augustine rejects the idea that the Spirit becomes incarnate as a dove at the baptism of Jesus," while acknowledging, nevertheless, that "it seems hard to escape the idea that the dove is at least an image of the Spirit at this point."[42] So the dove is an "image" but not an "incarnate" manifestation of Spirit? In spite of Clough's call for a theology of animal incarnation, I find his demurral on the question of the Holy Spirit's full-bodied enfleshment of God in Jesus' baptismal dove to be confusing.

I do not, however, want to overstate my critique of Hobgood-Oster, Allen, and Clough. Their silence or hesitancy to ascribe full animal identity to the Holy Spirit is understandable in light of many historical Christian thinkers' unwillingness to make a similar ascription. I believe the fear that underlies this unwillingness is the specter of pantheism that haunts all attempts to correlate corporality and divinity. But *Christian* animism is not pantheism—nor is it unadulterated animism per se.[43] On the contrary, the model of animism in a biblical register I am suggesting alternately sounds two different but complementary notes: the enfleshment of God in the world vis-à-vis Jesus' humanity and the Spirit's animality, on the one hand, and the alterity of God in God's self as heterogeneous to the world, on the other. Christian animism does not elide the differences between God and the world—as can happen in some pantheistic and animistic formulations of the God-world relationship—insofar as God and world are not collapsed into the same reality without remainder. Instead, it sets forth both the continuity and disparity between the divine life and earthly existence. The paradoxical logic at the heart of Christian animist grammar can be put, in philosophical terms, as dialectical monism or, in theological terms, as *coincidentia oppositorum*. By creating all things in the divine image, by becoming human and animal flesh and living among us, by pouring Godself out into the world, God is wholly "the same" as us. But God in God's ineffable and unknowable mystery—indeed, is this not the meaning of the crucifixion wherein God, the "courageous God who dares to commit suicide,"[44] traumatizes Godself by abandoning God's son in the moment of Jesus' cry of dereliction on the cross, "My God, my God, Why have you forsaken me?"—is also "other" from the world and, at times, or so it seems, strangely divorced from mortal affairs, human or otherwise.

Otherness and sameness. Unity and multiplicity. Transcendence and immanence. Aseity and kenosis. Contemporary theologians have used a variety of paradoxical phrases—"nonoppositional dualism,"[45] "sacramental embodiment,"[46] "apophatic entanglement"[47]—to articulate the aporia of God's alterity and inseparability from creation. Their point is that the world is a continuous self-expression of divinity with no a priori restrictions attaching to this self-expression.

In the Chalcedonian christology of early creedal Christianity, the humanity and divinity of Christ fully circulate together in one person without confusion or separation. In turn, this grammatical formula generates the theological syntax for parsing the omnierotic relationship between God and the world: without division, both now intertwine each other in unbounded love and intimacy, but without any confusion in the identities of the two distinctive orders of being.

Martin Buber makes this point well by poetically spelling out the restless longing both God and humankind have for each other. Buber says, "That you need God more than anything, you know at all times in your heart. But don't you know also that God needs you—in the fullness of his eternity, you? How would man exist if God did not need him, and how would you exist? You need God in order to be, and God needs you—for that which is the meaning of your life."[48] For Buber, God and ourselves—and I would expand this God-human relationship, as Buber himself does, to include God and *all* beings—share a common longing for fulfillment in the being of the other. Softly with feeling: opposites fuse into a tensive parity with each other. Softly with feeling: polarities flow into a differentiated unity between the one and the other. Softly with feeling: God (erotically charged) and the world (achingly amorous) caress each other in mutual attraction and filiation. I regard Buber's model of God's passionate need for a mutually affirming relationship with others, which is neither pantheist nor reductively animist, to be deeply resonant with the model of commensality between God and the Earth in Christian animism.

In sum, this book turns on a simple but I hope groundbreaking question. Could it be that Christian faith, at its core, centers on belief in God as a fully incarnated reality not only in the humanity of Jesus Christ but also in the animality of the Holy Spirit, even though this core insight has rarely been recognized as central to Christian identity? Could it be that the basic system of Christian belief is founded on a permeable and viscous God becoming not only human flesh in the person of Jesus but also animal flesh in the person of the Spirit and that, if this is the case, is not the wide-ranging world of nonhuman nature—the birds of the air, the fish of the sea, the beasts of the field, the trees of the forest—the focus of God's interest, not just human well-being? And if this is the case, should we not, as human beings, comport ourselves toward the natural world in a loving and protective manner because this world is the fullness of God within the life of every creature?

To this end, *When God Was a Bird* weaves together biblical interpretation, historical theology, philosophical analysis, and my own nature writing in a tapestry of Christian animist vision. In telling my story, I am inspired by the American naturalist and theologian John Muir, whose larger-than-life narratives of wild nature, as Brian Treanor puts it, have "successfully induced many people to value and, after a fashion, to love places that they themselves would in all likelihood never see or experience in person. . . . John Muir's narratives about Yosemite and the High Sierra *did* induce people to love, and consequently preserve, these natural treasures."[49] Relying on personal narratives underscores my practical aim in this book: to inculcate in readers a deep feeling of belonging with our terrestrial kinfolk so that we will want to nurture and care for them as common members of the same family. And in order to give life to this model of kinship relations I envision here, I preface each chapter with an original woodcut illustration by the contemporary artist James Larson of particular wild birds that I or, in one case, John Muir have met in our journeys within Earth's sacred landscapes. I offer these drawings as testimonies to the numinous wonder of commonplace birds—feathered traces of divinity—within daily existence.

I conclude with a précis of the book's overall structure. Chapter 1 opens with my encounter with the song of the wood thrush and then focuses on divine animals in the Bible. It examines the Gospels' "pigeon God," in which the Spirit-bird alights on Jesus at the time of his baptism, signaling the unity of all things: divine life and bird life, divinity and corporeality, spirit and flesh. And it argues that the Bible's seeming prohibitions against animal deities is vitiated by Moses' and Jesus' ophidian *shamanism*, which privileges snake totemism as a source of salvation in the book of Numbers and the Gospel according to John, respectively. It examines intimations of Christian animism—the belief that all things, including so-called inanimate objects, are alive with sacred presence—in George E. "Tink" Taylor, Lynn White Jr., and the *Martyrdom of Polycarp*, a second-century CE avian-spirit-possession narrative. I conclude that insofar as the Spirit is *ornithomorphic*, it behooves us to care for the natural world as the site of God's daily presence.

Chapter 2 begins with me and my students taking a hydraulic fracturing ("fracking") tour of northern Pennsylvania to witness the devastation wrought by extreme energy extraction. In Martin Heidegger, this type of technology is an exploitative "setting-upon" nature, rather than "bringing-forth" nature's latent possibilities in a manner that is site appropriate and organic. Healthy interactions with nature are resonant with the "incantatory gesture" characteristic of Christian animism: summoning the presence of the numinous within the everyday. Glossing Mary Douglas and Julia Kristeva, this chapter shows that Jesus, the good shaman, is a model of "bringing-forth" when he mixes saliva and dirt together in a poultice to heal the blind man in John 9. According to René Girard, however, nature is a site not of healing but of dangerous boundary violations. The chapter concludes with a vignette about my viewing the pileated woodpecker, sometimes called the "Lord God!" bird by awestruck onlookers, in the Crum Woods near Swarthmore College. Like the aerial Spirit at Jesus' baptism, I suggest that catching sight of this avian deity reconciles the two orders of being—*divinity* and *animality*—that Girard seeks to drive apart.

Chapter 3 starts with a *visitation* by a great blue heron to my Religion and Ecology class taught in the Crum Woods. Is the Crum Woods *holy ground*? Some ecotheologians (John B. Cobb Jr., Richard Bauckham) caution against this way of speaking, but I propose that Christianity is a religion of *double incarnation*: in a twofold movement, God becomes flesh in both humankind (Jesus) and otherkind (Spirit), underscoring that the supernal and the carnal are one. The chapter focuses on historical portraits of Jesus' relationship to particular birds (sparrows, ravens, and roosters) as *totem beings* in his teaching ministry; Augustine's repudiation of Neoplatonism and his natalist celebration of the maternal, birdy Holy Spirit in the world; and Hildegard of Bingen's avian pneumatology, in which Earth's "vital greenness" is valorized for its curative powers in a manner similar to Jesus' mud-pie healing of the blind man in John 9, noted in Chapter 2. I conclude with a meditation on *nature worship* as acceptable Christian practice in a Quaker meetinghouse in Monteverde, Costa Rica.

Chapter 4 keys on John Muir's ecstatic wilderness religion as a paradigm of the dialectic between Christianity and animism at the heart of this book, namely, *Christianimism*. Muir's nature evangelism, however, came at the price of rhetorically abetting the forced removal of Native Americans from their homes within the fledgling national parks movement, including Yosemite National Park. Notwithstanding this stain on Muir's legacy, his thought is notable for rethinking the full arc of Jesus' life—baptism by John, departure into wilderness, walking on the water, throwing out temple money changers, farewell discourses, and crucifixion—vis-à-vis his own life in terms that are deeply personal as well as being environmentally and biblically sonorous. Glossing Northrop Frye, Muir's artful use of the Bible is the *great code* that unlocks his euphoric nature mysticism. Some contemporary interpreters of Muir miss this point (Michael P. Cohen, Bron Taylor), but Muir advocates a *two-books theology* in which the Bible and the Earth are equally compelling revelatory "texts." His *Yosemite spirituality* reaches its apogee in his 1870 "woody gospel letter," a paean to a homophilic, orgasmic religion of sensual delight: "Come suck Sequoia and be saved." In Muir's spirit, I conclude today that Christianity is still not Christianity—in spite of its deep incarnational grammar—because of its sometime hostility to embodied, earthly existence.

Using James Lovelock's *Gaia* theory and current biblical exegesis, Chapter 5 maintains that Earth is a *sentient organism* with its own emotional registries, relational capacities, and vulnerability to suffering. This "living Earth" theme is further explored in case studies of two sacred land sites in northern Spain that my wife, Audrey, and I traveled on foot: the Cape of the Crosses national park and the El Camino de Santiago pilgrimage route. I encountered both sites as "thin places"—landscapes where divinity and materiality comfortably intersect—in which errant wandering and purposeful travel were valued equally. But in our current state of social and environmental inequity, such sites are also agonizingly *cruciform*: as Jesus was sacrificed at Calvary, so today we crucify afresh God's winged Spirit in nature through toxic impacts against the very life-support systems that make all beings' planetary existence possible. Massive species depredation—iconically signified by the extinction of the passenger pigeon, which used to soar in great flocks across American and European woodlands—provokes the question of whether God's presence can still be felt when traces of avian divinity are being wiped out, "taking our feelings" with them, in the indictment over the poet W. S. Merwin. The scars of Golgotha mark the whole Earth. The wounds of crucifixion extend on all of creation. The book concludes on a note of broken hope symbolized by the feral pigeon—the dovey cousin of the passenger pigeon and also of Jesus' baptismal bird that Audrey and I witnessed again at the end of our El Camino trek—amid the contemporary loss of embodied deity through *ecocidal*, even *deicidal*, practices.

CHAPTER I

# Song of the Wood Thrush

## *The Singing Monk of the Crum Woods*

Today the wood thrush returned to the Crum Woods.[1] I have been waiting for this event for months. I moved to a house in the forest five years ago, and at that time, I heard a strange and wonderful bird call in the tree canopy. The song of the wood thrush is a melody unlike anything I had ever heard. Liquid, flute-like, perfectly pitched—the thrush vocalizes a kind of duet with itself in which it simultaneously produces two independent musical notes that reverberate with each other. I have read that Tibetan monks can also sing two notes at the same time, a baseline and a melody line in contrapuntal balance, so now I think of the wood thrush as the singing monk of the Crum Woods.[2]

In the spring and summer, I wake up, and often go to sleep, to the vocal pleasures of a bird that I cannot see but whose delicate harmonies pleasantly haunt my dreams. Like God's Spirit, I know the thrush is there—I hear its lilting cadence from dawn to dusk—but I've only seen one wood thrush during the time I've lived in the Crum Woods. I creep around the forest floor looking skyward, hoping for a sighting, but it always escapes my gaze. Instead, I keep my window open at night as a vector for the thrush's call. Bathed in its music, it is hard for me to distinguish between waking and sleeping, between twilight, midnight, and early morning. At dusk, the thrush is in my ear until I fall asleep; I dream of its call throughout the night; and I wake up after dawn gently moving through the deep of its sweet-sounding counterpoint.

The wood thrush lives in the interior of the Crum Woods and consistently refuses the lure of my feeder. Thrushes prefer just the right habitat blend for sustenance and breeding: running water, dense understory cover, and moist healthy soil full of fruiting plants and insects to eat. In the heart of the forest, foraging in the leaf litter among large deciduous trees, the thrush makes its nest out of dead leaves, mud, twigs, and sometimes found manufactured materials such as paper and plastic. Like other neotropical songbirds, it is threatened by habitat loss through continued development of its home range. It is also endangered by brood parasites, such as brown-headed cowbirds, which lay their own eggs in wood thrush nests, crowding out the host's eggs and hatchlings. The perdurability of the thrush in the face of these obstacles gives me hope in a time of despair about the world's future. Henry David Thoreau says, "The thrush alone declares the immortal wealth and vigor that is in the forest. . . . Whenever a man hears it he is young, and Nature is in her spring; wherever he hears it, it is a new world and a free country, and the gates of Heaven are not shut against him."[3] For me, Earth and heaven come alive with mystery and wonder when I hear the thrush's ethereal song. In my own particular bioregion, the thrush opens to me the beauty of the Crum Woods as a vital habitat—indeed, as a sacred forest—whenever I am graced by its stirring music.

## *Nature Religion*

To call the Crum Woods a *sacred forest* may seem odd if one is using traditional Christian vocabulary.[4] Historically, many Christian thinkers avoided ascribing religious value to natural places and living things and restricted terms such as *sacred*, *holy*, and *blessed* to God alone. Central strains of classical Christian opinion desacralized nature by divesting it of religious significance. While the Bible is suffused with images of sacred nature—God formed Adam and Eve from the dust of the ground; called to Moses through a burning bush; spoke through Balaam's donkey; arrested Job's attention in a whirlwind; used a great fish to send Jonah a message; and appeared alternately as a man, a lamb, and a dove throughout the

this seems the only necessary argument

New Testament—Christianity, in the main, evolved into a sky-God religion in which God was seen as an invisible, heavenly being not of the same essence as plants, animals, rivers, and mountains. As the theologian E. O. James writes, the Christian "God of heaven . . . was always regarded as transcendentally distinct from the rest of creation, physical, human, biological. . . . He was the Sky-god *par excellence* because, in scholastic terminology, He was 'He Who is.'"[5] The Welsh "St. Denio" hymn I sing in my church proclaims the same sentiment: "Immortal, invisible, God only wise, in light inaccessible hid from our eyes." Hidden and imperceptible, God exists in a far-removed place divorced from the ebb and flow of mortal life here on Earth.[6] Moreover, God the creator alone is holy, so goes this line of thinking, and everything else in creation, derivatively made by God, is an extension of God's blessed and benevolent handiwork—but not independently good and holy unto itself.

But in the Earth-centered narrative arc of the biblical stories I recover in this book, this devaluation of nature as devoid of sacred worth is conspicuously absent. Even at the outset of the Bible, God is not an invisible sky-God "transcendentally distinct from the rest of creation, physical, human, biological," as E. O. James avers, but a fully incarnated being who walks and talks in human form, sprouts leaves and grows roots in the good soil of creation, and—clothed in feathers and flesh—takes flight and soars through the updrafts of wind and sky. An astoundingly rich variety of natural phenomena are charged with divine presence in the biblical accounts, with God appearing alternately in human and plant forms—and in animal form, as I will highlight here.

To this end, let us start with the winged bird-God of creation, the central figure in the Bible's inaugural creation story. In the beginning, the Earth was formless and empty, and God's Spirit swept across the dark waters of the great oceans. "In the beginning, God created the heavens and the earth. The earth was formless and void and darkness covered the waters. And the Spirit of God [*rûach elohim* in Hebrew] hovered [*merahefet*] over the face of the deep" (Genesis 1:1–2). The noun *rûach elohim* that is used by the Genesis authors to identify Spirit is grammatically feminine, while the feminine-ending verb form that is employed to describe the Spirit's movement is *merahefet*, alternately translated as to "hover over," "sweep over," "move over," "flutter over," or "tremble over." This feminine, avian noun-verb cluster describes the activity of a mother bird in the care of her young in the nest. One grammatical clue to the meaning of this dynamic expression can be found in Deuteronomy 32:11, where God is said to be a protector of Jacob in a manner akin to how "an eagle stirs up its nest, and hovers [*merahefet*] over its young."[7] Using the same winged imagery deployed by the Deuteronomic author, the writer of Genesis describes the Spirit as a flying, feathered being—a bird or something like a bird—to describe its nurturing care over the great expanse (perhaps we should say the great egg?) of creation. Analogous to a mother eagle brooding over her nest, God's avian Spirit, hovering over the face of the watery deep, is a "giant mother bird,"[8] a divine-animal hybrid that challenges the conventional separation of the divine order and the animal kingdom in much of classical Christian thought.

Another clue to God's fleshly identity is found in Genesis 32. In this story, Isaac's son Jacob is seeking to enter the land of Canaan, the biblical promised land, and reconciliation with his brother, Esau. On the evening before these events, Jacob is alone at the Jabbok River, and there he encounters a strange nocturnal visitor who demands a wrestling match with him, which lasts all night. Neither man can subdue the other, but the night wrangler, vampire-like, eventually cries out to be let go because the first morning light is about to appear. Jacob then insists he must be blessed first, and his opponent accedes to his request, changing Jacob's name to Israel. At this point, Jacob also does a name switch, calling the place of struggle "The Face of God" (*peniel* in Hebrew) because here, on the banks of the river, he said, "I have seen God face to face, and yet my life is preserved" (Genesis 32:30). As in the previous example, God again is creaturely, but now as a man, not a bird. Yet the meaning is clear: Jacob sees God directly in his face-to-face encounter with the mysterious night brawler—and not only survives his altercation with him but is blessed in the process.

A further example of God as a corporeal life-form is Exodus 3, the story of the burning bush at Mount Horeb (or Sinai). Here Moses is surprised to hear God speaking though a bush (*seneh* in Hebrew, meaning "bramble" or "branches") that is aflame with a nonconsumptive fire. Because God speaks in and through the bush itself, the story shows how God implants Godself as a fiery bramble but is somehow not consumed in the process of being on fire. As the holy fire plant, God tells Moses, "Do not come near; take off your shoes from your feet because the place on which you are standing is holy ground. . . . And Moses hid his face because he was afraid to look at God" (Exodus 3:5–6). In this account, inviolate boundaries between humankind, plant life, and divinity are crossed. The story erases time-honored binary oppositions about who or what God is, and where God is located, in this mix of scrubby fire, commanding voice, and dangerous ground. Again we confront the shock of viscous carnality as Moses hides his countenance because he is afraid to see God face-to-face—God the sacred plant, God the consecrated ground, God the burning fire. Speaking to Moses in a commanding voice on the holy mountain, God in Exodus is an on-fire, earth-rooted vegetation deity (*deus botanicus*).[9]

What strikes me as common in all three of these stories is that God—as avian, human, and vegetal, respectively—is biologically elemental. Provocatively, in these various forms of God's feral bodies, God is feathers and bone, blood and skin, leaves and wood. We think we know whom or what God is—for Jews, Torah; for Christians, Jesus—but God often presents Godself to us in somatic forms of life that surprise and challenge our common assumptions. This question, therefore, is the driving question of this book: If feathers, bone, blood, and wood constitute God in the Bible, could not these same wild elements be God's living presence among us today? If God was the creation bird in Genesis, the night brawler in the Jacob story, and the fiery thicket in Exodus, then could not the birds, people, and plants among us today be God-in-the-flesh once again? Is it possible to reenvision these ancient biblical life-forms—spirit bird, divine man, sacred bush—as precursors of God's corporeal presence within our *own* everyday natural existence in the present? And if this is the case, then should we not relate to all things—the birds of the air; every person we encounter; and the brambles, trees, and flowers that grace our existence—as, once again, the miraculously animated manifestations of the birdy, fleshy, leafy God of the biblical witness? If all creation is *sacred*, in other words, should we not comport ourselves to the natural world with reverence and adulation as the enfleshment of God in the biosphere?

Mother Theresa often spoke of finding God in everyone to whom she ministered, that the "unloved are Jesus in disguise."[10] Expanding her comments beyond the human to the more-than-human world, could it be that all of the things that we do not value as bearers of the sacred, the "unloved" as it were, that all of what we take for granted within profane, everyday existence, that all of this is God in God's many wonderful and sometimes terrifying disguises, that the great expanse of the natural world and all of its many denizens *is* who and what and where God is today? Could it be possible that God is porous and hidden under the veil of common life? That all of the good things God creates—the verdant myriad of our biotic and abiotic kinsfolk who walk, fly, run, creep, flow, and grow roots on, through and in the Earth—are all divine, holy things, now living in our midst in all of God's various and sundry disguises? If the core grammar of Christianity is that God became flesh and dwelt among us

(John 1:14), can we not say, then, that all flesh is sacred, that all nature is blessed, that all things are holy? And if we can say this, is not the Christian religion an exuberant testimony to this spiritual truth, namely, that the sacred is the profane, the heavenly the worldly, the spiritual the earthly, and the godly the carnal?

To be sure, to suggest that all of creation and its many inhabitants are God in a variety of forms and disguises is a difficult notion to swallow for some people of faith. As Karl Barth says, "God may speak to us through Russian Communism, a flute concerto, a blossoming shrub, or a dead dog. We do well to listen to him if he really does."[11] But some Christians find such latitudinarian attitudes excessive. I recently gave a talk to a nearby church group in which I interpreted the biblical stories of the Holy Spirit as an avian form of divinity. I used the account of the bird-God of Genesis noted earlier and said the same imagery of God on the wing is used in all of the Gospels as well (a point I will expand on shortly). Could the dovey Spirit of Christian witness who appears at Jesus' baptism—fluttering over Jesus in the Jordan River in a manner analogous to the hovering Spirit bird in Genesis—signal that as God became flesh and feathers in biblical time, God could become flesh and feathers in our own time as well? Could this mean, I asked, that the biblical revelation of the avian God of the Bible entails that God could appear again today in the form of a bird or, in principle, any other life-form? At this point, a member of the audience raised her hand and spoke up: "My brother doesn't like doves. He has mourning doves in his yard. He doesn't like their whistling when they fly. In the morning they make too much noise, so he gets up and shoots them wherever he finds them." Speechless, it was clear that my attempt to make a case for God's full and promiscuous incarnation within the natural world did not make sense according to the deep faith shared by at least some members of this church community.

## *The Pigeon God*

I have suggested, biblically speaking, that an astoundingly rich variety of quotidian phenomena are charged with divinity—the world is a fecund and luxuriant riot of God's presence—including the bodies of sacred animals, such as the Genesis bird-God. Presumably a critic of a theology of divine animality would object to my interpretation of Genesis 1 as a matter-of-fact description of the Spirit bird in creation as God on the wing and counter that the Genesis account is a figure of speech. Such a critic would likely maintain that the creation bird is a rhetorical device, a birdy metaphor for articulating God's all-pervasive presence in creation, like a hen over her nest, poetically expressed, but not an actual description of God's winged body. But when the creation account is read in tandem with the New Testament's similar descriptions of the airborne Holy Spirit, what emerges is a thoroughgoing biblical pattern of depicting God's Spirit in avian terms.

In the story of Jesus' baptism in the four regular Gospels—and in one unauthorized Gospel called *The Gospel of the Ebionites*, an early-second-century CE narrative about Jesus that harmonizes the other canonical Gospels into one account[12]—the Spirit, much like in the Genesis account, comes down from heaven as a bird and then alights on Jesus' newly baptized body. All five accounts narrate the same gospel memory, namely, that as Jesus presents himself to be baptized by John the Baptist and is baptized, the Spirit descends on Jesus as a dove from heaven, and then, in the synoptic and extracanonical Gospels, a voice from heaven says, "This is my beloved son with whom I am well pleased." At the time of Jesus' baptism, it seems certain that the power and wonder of the descending Spirit-bird, along with a heavenly voice, indelibly seared the memories of the authors of each of the five Gospels. This collective memory of the feathered divine being appearing at the debut of Jesus' public ministry must have etched a lasting image in the minds of each of the canonical and extracanonical authors, all of whom told, roughly speaking, the same story. Here is the record of the five related accounts of Jesus' baptism and the winged Spirit:

> Then Jesus came from Galilee to John at the Jordan, to be baptized by him. John would have prevented him, saying, "I need to be baptized by you, and do you come to me?" But Jesus answered him, "Let it be so now; for it is proper for us in this way to fulfill all righteousness." Then he consented. And when Jesus had been baptized, just as he came up from the water, suddenly the heavens were opened to him and he saw the Spirit of God descending like a dove and alighting on him. And a voice from heaven said, "This is my Son, the Beloved, with whom I am well pleased." (Matthew 3:13–17)
>
> In those days Jesus came from Nazareth of Galilee and was baptized by John in the Jordan. And just as he was coming up out of the water, he saw the heavens torn apart and the Spirit descending like a dove on him. And a voice came from heaven, "You are my Son, the Beloved; with you I am well pleased." (Mark 1:9–11)
>
> Now when all the people were baptized, and when Jesus also had been baptized and was praying, the heaven was opened, and the Holy Spirit descended upon him in bodily form as a dove. And a voice

came from heaven, "You are my Son, the Beloved; with you I am well pleased." (Luke 3:21–22)

And John testified, "I saw the Spirit descending from heaven like a dove, and it remained on him. I myself did not know him, but the one who sent me to baptize with water said to me, 'He on whom you see the Spirit descend and remain is the one who baptizes with the Holy Spirit.' And I myself have seen and have testified that this is the Son of God." (John 1:32–34)

When the people were baptized, Jesus also came and was baptized by John. When he came up out of the water, the heavens opened and he saw the Holy Spirit in the form of a dove, descending and entering him. And a voice came from heaven, "You are my beloved Son, in you I am well pleased." (*The Gospel of the Ebionites* 4)

On one level, I suspect that the people who came to John for baptism were not surprised to see the Holy Spirit in the form of a dove. In biblical times, doves—in addition to the other divinized flora and fauna I noted earlier—figured prominently in the history of Israel as archetypes of God's compassion. Noah sends a dove out after the flood to test whether dry land has appeared (Genesis 8:6–12). Abraham sacrifices a dove to God to honor God's covenant with him to make Israel a great nation (Genesis 15). Solomon calls his beloved "my dove," a heartfelt term of longing and endearment (Song of Solomon 2:14, 4:1, 5:2, 6:9). And Jeremiah and Ezekiel refer to doves' swift flight, careful nesting, and plaintive cooing as metaphors for human beings' pursuit of nurture and safety in times of turmoil and distress (Ezekiel 7:16; Jeremiah 48:28). As divine emissary and guardian of sacred order, the dove is a living embodiment of God's protection, healing, and love.

On another level, however, Jesus' audience at the time of his baptism must have wondered, is God the Spirit really a beaked-and-feathered creature? In all five of our texts, the Greek term used to denote this winged being is *peristera* (*Columba livia* in Latin). Interestingly, in Greek, as in English, the name for this sort of bird can mean either "dove" or "pigeon," a confusion that is papered over by the English New Testament translations of Jesus' baptismal bird as a dove, as well as by later artistic depictions of this bird as a beautiful and gently ethereal white dove.[13] But the original Greek word for this creature—*peristera*—has a wider semantic range and can mean either one of these two types of birds, dove or pigeon.[14] In English taxonomy as well, both the dove and the pigeon constitute the same "branch" on the avian "tree" that ornithologists use to classify birds of all types.

In today's common terminology, this one bird with different names is alternately referred to as the *rock dove*, the *rock pigeon*, or the *common city pigeon*. All of these names refer to the same bird, sharing different variations in plumage. Historically, scholars think the rock dove was the aboriginal form of this bird and as likely the iconic heavenly dove referred to in the Gospel texts in question. Today, the rock dove, rock pigeon, city pigeon, or what most people simply call "the pigeon" can be found on every continent (except Antarctica), many oceanic islands, and almost every habitat strewn across the Earth. It is as likely, then, that God's avian Spirit in the Gospels was as much a rambunctious feral pigeon as it was a virginal all-white dove. The pastor and theologian Debbie Blue makes this point in her excellent study of birds in the Bible:

> A dove is a pigeon. That seems worth saying repeatedly. We may have imagined that the dove at [Jesus'] baptism was white, but it was more likely gray, with an iridescent green-and-violet neck—a rock dove, which is very common in Palestine and which is considered to be the ancestor of our common domestic pigeon—the kind that gathers in our parks, nests in our eaves, poops all over buildings and sidewalks.
>
> The rock pigeons found in our cities and barns are probably from populations established by escaped domestic pigeons. They are often referred to as feral pigeons. How is that for a symbol of the Holy Spirit? I believe it's a good one. I like it. It's ubiquitous, on the streets. The white dove is overused. How about pigeons for Pentecost, on banners and bulletin covers?[15]

So the question remains, is God really a bird? To be specific, Is God a grayish, greenish, purple-colored domestic or feral pigeon who lives off seeds and berries, coos and whistles in flight, provides good eating for humans and other animals, and mates for life? Historically, most Christians have expressed confidence in the idea of God as Trinity, that God is both one in unity with Godself and expressive of Godself in three persons or modes of being, namely, Father, Son, and Holy Spirit. While Father and Son are often depicted in human form (God is both heavenly parent and the son sent from heaven to earth), the Spirit is generally figured in animal (dove/pigeon) and elemental (earth, fire, water, air) terms. By way of analyzing the Spirit's avian nature, let us briefly consider the Spirit's elemental character. In biblical terms, it becomes immediately clear that the four cardinal elements that make up all of life—air, water, fire, and earth—are the same fundamentals that constitute the biblical Holy Spirit as well.

First, the Spirit is *air*. When God breathes air into the lungs of the first human being ("then the Lord God formed adam from the dust of the ground, and breathed into his nostrils the breath of life, and adam became a living being"; Genesis 2:7), the Hebrew word for this primordial breath of life (*rûach*, or *pneuma* in Greek) also means Spirit. God's Spirit is God's breath, signaling that the atmosphere itself, all that we and all other beings need for survival, is the animating power of divinity in our midst—God swirling and blowing around us, making all things live.

Second, the Spirit is *water*. Indeed, Jesus associates the Spirit with living water ("Let the one who believes in me drink. As the scripture has said, 'Out of the believer's heart shall flow rivers of living water.' Now he said this about the Spirit"; John 7:38–39). This association signals that the Spirit is not only pneumatic but also aquatic: God's Spirit is the watery landscape—rivers, seas, vernal springs, life-giving rain—that makes planetary existence robust and sustainable.

Third, the Spirit is *fire*. After Jesus' ascension, tongues of fire alight on his followers (perhaps an allusion to the nonconsumptive fire of the burning bush), and without being burned by the fire, Jesus' followers begin to speak in different languages ("And suddenly from heaven, there came . . . tongues of fire, and a tongue rested on each of them. All of them were filled with the Holy Spirit"; Acts 2:2–3). Fire can destroy, but it can also provide warmth and comfort; here the disciples and others are heartened because each can hear the others speaking in their native tongue aided by the fiery Spirit of God.

Fourth, the Spirit is *earth*. We have seen the powerful association between Earth or earthen beings such as the bird-God in, for example, the story of creation, where the Spirit hovers over the great expanse of the planet, uniting inseparably God and the world ("In the beginning God created the heavens and the earth . . . and the Spirit of God hovered over the face of the deep"; Genesis 1:1–2).

It is this particular bird-like image that I want to focus on in Jesus' baptism stories in the Gospels—and especially one of the Gospels, Luke. Luke's story of Jesus' baptism, and concomitant presentation of the bird-God, is a thoughtful summary of the Gospels' overall narrative of Jesus' ritual immersion. This is Luke's account in the Revised Standard Version of the Bible: "Now when all the people were baptized, and when Jesus also had been baptized and was praying, the heaven was opened, and the Holy Spirit descended upon him in bodily form, as a dove, and a voice came from heaven, 'Thou art my beloved Son; with thee I am well pleased'" (Luke 3:21–22). After highlighting Jesus' baptism by John and then the opening of the heavens, Luke says (with notation of the original Greek text), "the Holy Spirit descended upon [Jesus] in bodily form [*somatiko eidei*], as a dove/pigeon [*hos peristeran*]" (Luke 3:22). Here the Greek phrase *somatiko eidei* means "in bodily form" or "in bodily essence." In this phrase, the Greek adjective *somatikos*, from the noun *soma* (body), signifies the shape or appearance of something in corporeal form. The point here is that God as Spirit is fully carnal and fully creaturely. Moreover, God as Spirit is fully creaturely in the form of an animal body, the dove/pigeon of the Gospel witness. Now the Holy Spirit, the third member of the Godhead, enters existence in *animal* form—even as the second member of the Godhead, Jesus, enters existence in *human* form.

In all five of the Gospel baptism stories, God as Spirit becomes a very specific type of animated physical body: a seed-eating, nest-building, air-flying member of the dove/pigeon order of things. The particular beak-and-feathers body Luke's spirit-animal becomes is defined by the phrase *hos peristeran*, which means "as a dove/pigeon," "even like a dove/pigeon," or "just as a dove/pigeon"—that is, the Spirit's body is thoroughly bird-like. Some English translations of the Lukan and other Gospel accounts of Jesus' baptism miss this point. While the Revised Standard Version says, "The Holy Spirit descended upon him *as* a dove," the New Revised Standard Version prefers, "The Holy Spirit descended upon him *like* a dove" (emphases added).

But the preposition *hos*—from *hos peristeran* in the original Greek text of Luke 3:22 and elsewhere—does not operate here metaphorically or analogically but appositionally. The expression "as a dove/pigeon [*hos peristeran*]" in this context is not a simile that says that the Spirit descended in bodily form *like* a dove/pigeon but is rather an appositional phrase that qualifies the meaning of the *actual* physical being the Spirit has become. In other words, the Spirit descended in bodily form *as* a dove/pigeon—or in the technical language of the history of religions scholar Mircea Eliade, the Spirit is *ornithomorphic*.[16] In the grammar of predication, the Spirit *is* a dove/pigeon, not *like* a dove/pigeon. Luke 3:22 is not, then, a figure of speech to connote the passing bird-like appearance of the Spirit in this one instance but a literal description of the actual bird-God the Spirit is or has become.

Here in this discussion, I have alluded to parallels between Genesis's and the Gospels' portraits of the Holy Spirit. A final clue that ties together the Genesis account of the hovering avian divinity with the Gospels' bird-God is found in Matthew's and John's version of Jesus' baptism. All four Gospels narrate the heavenly descent of the dovey Spirit upon Jesus at the time of his baptism. But Matthew and John add key verbs to the scene that resonate with the ancient echoes of the Bible's inaugural creation hymn. Matthew says that the Spirit "lights upon" or "comes upon" Jesus (the Greek verb is *erchomai* in both cases), emerging from the Jordan River, on the one hand, while John writes that the Spirit "stays with," "abides with," or "remains with" (here the word is *meno*) Jesus during this scene, on the other. This vibrant sight of the Spirit in the Gospels both descending upon and then hovering about the newly baptized Jesus harks back to the Spirit in Genesis hovering or fluttering over the newly created Earth. In both stories, the creator Spirit *tarries* with her chosen subjects—Earth in the one case, Jesus in the other—in a gesture of benevolent warmth and concern.

In my mind's eye, the winged Spirit in the Genesis story, the glorious mother bird of creation, tenderly broods over the great egg of the cosmos in an attitude of sustained solicitation and affection. Likewise, in the Gospels, I also imagine the heavenly bird circling about Jesus' head protectively, cooing and whistling her approval of his divine identity and mission—even, perhaps, possessively perching on his shoulder, then hovering and flying about him, and maybe even building a nest in his hair, signaling tenderness and love toward Jesus, his body still glistening wet from the life-giving river as the sacred bird crowns his vocation with her attentive care. Gently alighting on Jesus' person, just like the creation bird hovering over the deep in Genesis, the Gospels' heaven-sent dovey pigeon is God enfleshing Godself in carnal form, but now not only in human flesh in the person of Jesus (God's Son) but also in animal flesh in the person of the Spirit (God's Spirit).

Theologically speaking, this winged God in the Hebrew Bible and the New Testament champions the eternal unity of all things: heaven and Earth, divine life and birdlife, divinity and animality, spirit and flesh. Could it be then, glossing the Nicene Creed, that the Holy Spirit, the one who with the Father and Son is worshipped and glorified, that this same Holy Spirit is a lowly, ordinary pigeon—so much so that when we look into the black pupils and brightly colored irises of a common pigeon, we are looking into the face of God, the Lord and giver of life, in the language of the Nicene Creed?

## *Sacred Animals*

In this book, the heart of my argument is that God in biblical times was encountered as a bird-God—and that this encounter opens up the possibility that all things today are filled with God and thereby deserving of our reverence and care. But at first blush, there are good reasons why my call for an avian model of God does not make sense. The examples I gave of God as a bird, man, and bush in the early biblical history, and especially as a bird in the Gospels, is not a convincing argument without careful study of what seems to be the Bible's definitive statement against religious naturalism: God's thunderous denunciation of the Israelites' worship of the golden calf at Mount Sinai. In biblical terms, this story appears to count against my case for divine animality. Is not the point of the story that God condemns Israel for worshipping the image of a young cow, rather than the one true God, who is invisible and noncorporeal?

The story of God's condemnation of false worship takes place in the aftermath of the exodus from Egypt, which some scholars date to the thirteenth century BCE. At this time, while the Israelites are blessed with the gift of the covenant tablets to Moses at Mount Sinai, Moses is delayed on the mountain. The Hebrew people grow tired of waiting for his return from Sinai, and ask Moses' brother, Aaron, to make a golden calf, saying, "These are your gods, O Israel, who brought you up out of the land of Egypt" (Exodus 32:4). Even though Aaron combines obeisance to the calf with a call for a festival to the Lord, using the same burnt offerings and sacrifices prescribed by the emerging Torah-based religious system, God becomes violently angry with the covenant people. In response, Moses, now down from the mountain, shatters the covenant tablets and burns into powder the golden calf. God, likewise, sends a plague on the survivors and, after initially threatening to destroy all of the freed Hebrews, still orchestrates the deaths of thousands of Israelites through internecine sacrifice.

The golden calf debacle is a searing moment at a pivotal point in the biblical narrative, and it becomes a source of ongoing anguish for Israel and the early church. In its aftermath, a sort of posttraumatic stress disorder seizes the religious imagination of Jews and Christians regarding the veneration of images that embody God in the everyday world. Deep suspicion lingers. After the golden calf, many attempts in the Bible to represent God's animated body in animal or physical form are roundly condemned. The evil king Jeroboam, for example, is excoriated by the prophets for setting up not one but two golden calves in a secessionist effort to divide the Israelite nation (1 Kings 12). Similarly, Paul, perhaps remembering the trauma of the golden calf stories, says in a famous sermon in Athens that God is "Lord of heaven and earth [and] does not live in shrines made by human hands. . . . [And] we ought not to think that the deity is like gold, or silver, or stone, an image formed by the art and imagination of mortals" (Acts 17:24, 29). In the agonizing wake of the golden calf, it seems impossible to imagine that the biblical God could in any way be identified with, or represented by, a creaturely life-form.

But the issue is not that simple. In fact, biblically speaking, there seems to be nothing wrong, in principle, with Aaron's casting of the golden calf—any more than Moses' decision, a few months later, to cast a bronze serpent (Numbers 21). In the same way that the Israelites provoked God by becoming impatient of waiting for Moses at Sinai, they also irritated God by growing tired of eating the same desert food after the exodus from Egypt. Moved to retribution, according to the book of Numbers, God sends poisonous snakes to bite and kill the freed people. The Hebrews then apologize to Moses for their complaints; and, in response, Moses molds out of bronze an image of a deadly snake, puts it on a pole, and says that any victims of a viperous attack can look to the sacred snake pole and live. "And the Lord said to Moses, 'Make a poisonous serpent, and set it on a pole; and everyone who is bitten shall look at it and live.' So Moses made a bronze serpent, and set it on a pole; and whenever a serpent bit someone, that person would look at the bronze serpent and live" (Numbers 21:8–9).

The biblical scholar Tikva Frymer-Kensky says the bronze serpent marks a moment of "rapprochement with God," in which the healing animal icon mediates God's love and presence to Israel. She writes, "The temple also contained a bronze serpent which was said to have been made by Moses at the time of the plague of snakes. Kept in the temple, this icon served as a focus for human desires for health and healing. The sacrificial cult itself was a powerful image of rapprochement with God: the sound, smell, and taste of the animals all served to remind the worshippers of the presence of God in their lives."[17] Moses uses the cast snake to direct sacred, healing energies to the Hebrew snake-bite victims—as an "icon," as Frymer-Kensky puts it, that serves "as a focus for human desires for health and healing." In a word, Mosaic snake religion is *shamanism*. Utilizing a type of primordial homeopathy, Moses treats "like with like" by harnessing an image of a serpent as a curative technique to overcome venomous snake wounds. Traditionally understood, as the religion scholar Michael York puts it, the "shaman's overall function is to ensure, maintain and/or restore his/her people's proper relationship with the natural environment and with the spiritual realm as it manifests itself through that environment."[18] The shaman is adept at maintaining communal and ecological balance by functioning as a conduit though which spiritual power is funneled to others. The shaman, then, ensures the social and biological well-being of the community through eco-medicinal rituals. In the Sinai wilderness, Moses expertly performs this shamanic role by venerating the molten serpent as the sacred animal medium through which healing divine power is channeled to his people.

Most religious history I have read divorces Judaism and Christianity from the study of primordial animist religions by isolating the ancient "savage" practices of shamanism, totemism, and animal worship from the "pure" monotheism of the so-called religions of the book. In this telling, Judaism and Christianity evolved away from the muddy heathenism of primitive religion. But in regard to Judaism, and the same could be said of Christianity, the history of religions scholar Howard Eilberg-Schwartz argues the following instead: "What formerly were construed as traits of savage traditions are now discovered at the heart of Judaism. Totemism, for example, is no longer just a phenomenon of primitive religions. Israelite religion had its own form of totemism. . . . As in 'primitive' societies, animals provided the foundational metaphors through which Israelites articulated their understanding of who they were and what they wanted to be. These metaphors, which provided an idiom for theological, national, and social reflections, fundamentally shaped the practices of Israelite religion."[19] Situating Moses in the history of religions, it makes sense, as Eilberg-Schwartz puts it, that "Israelite religion had its own form of totemism"[20] or, as Mircea Eliade puts it, that Moses deploys the snake as a "helping spirit" who mediates power to Israel.[21] Moses' ability to direct healing power through this helping spirit in animal form is a special skill set peculiar to shamans. The bronze serpent is Moses' "totem" or "spirit-animal," the nonhuman life-form that enables him to both access and funnel well-being to his followers in the desert.

In the Bible, snake shamanism continued long after Moses' healing ceremony, so much so that six hundred years after Sinai, the Israelites are still paying obeisance to the bronze serpent. At this time, in the eighth century BCE, an internal dispute arises between Israel's leaders concerning the propriety of public veneration of Moses' original snake totem, and one particular ruler, King Hezekiah, tries to put an end to the snake cultus once and for all. "He removed the high places, and broke the pillars, and cut down the sacred pole. And he broke in pieces the bronze serpent that Moses had made, for until those days the people of Israel had burned incense to it; it was called Nehushtan" (2 Kings 18:5). Be this as it may, it is important to note that more than half a millennium after Moses—and throughout the period of Joshua, David, Solomon, and Elijah—God's people are still paying homage to a cast-bronze serpent in the Jerusalem Temple, notwithstanding Hezekiah's eventual objections to this nature-based practice. Indeed, according to Frymer-Kensky, Hezekiah's so-called reforms did not completely purge Israel of its veneration of animals and other beings "on every lofty hill and under every green tree."[22] Frymer-Kensky continues,

> Despite the vigorous opposition of the prophets, the people of Israel maintained these rituals and did not find them incompatible with the worship of YHWH. In fact, the complex of altar, tree, hill, and megalith that characterized this worship was an ancient and integral part of Israel's religious life, and the "reforms" of Hezekiah and Josiah that destroyed this complex were a radical innovation rather than a return to some pristine purity. The tenacity of this worship may indicate its importance to the people of Israel; indeed, this nature-oriented worship may have enabled the people of Israel to feel the immanence of God and to continue to worship the abstract and demanding YHWH.[23]

Even though Hezekiah takes down the decorative temple serpent, "nature-oriented worship," as Frymer-Kensky writes, remains inextricably interwoven with the religion of Israel, because such worship "enabled the people of Israel to feel the immanence of God and to continue to worship the abstract and demanding YHWH." Biblical animal religion ensured God's intimacy and nurturance of the covenant people. In this way, the Israelites continuously oscillated between worshipping God in nature using, among other things, images of different animal beings, on the one hand, and worshipping God outside of the natural world without the assurance of animal divinities to comfort and reassure the community, on the other.

In spite of Hezekiah's iconoclasm, this oscillation between nature-affirming and nature-denying beliefs and practices continues long past the eighth century BCE. Even eight hundred years later, we see in the time of the Christian scriptures that Moses' reptile religion is still strong in Judea—perhaps not in actual religious practice but certainly, we learn, in the religious imagination of the Hebrew people. Indeed, the restoration of the serpent cultus becomes central in Jesus' ministry.

Atavistically, Jesus invokes the power of Moses' medicinal snake when he compares his own saving death to the salvation wrought by the serpent pole at Sinai. Jesus says, "And just as Moses lifted up the serpent in the wilderness, so must the Son of Man be lifted up, that whoever believes in him may have eternal life" (John 3:14–15). But while Jesus continues Moses' snake practice, there is an important difference between the two:

whereas Moses' bronze totem is a built object, Jesus' totemism is bound together with his own identity as healer and Messiah. Unlike Moses, Jesus himself has animistically shape-shifted into becoming the sacred serpent for the renewal of the people. In spite of Hezekiah's attempted take-down of the spirit-animal, Moses' viper religion lives on in Jesus' own direct identification with the gilded serpent. Jesus' animal shamanism is a powerful affirmation of the original healing and restorative power of Moses' totem. Hezekiah's attack on the Mosaic snake cultus notwithstanding, the Bible articulates an unbroken line from Moses to the reign of David and the time of Jesus as a continuous exercise in shamanic healing featuring one of God's nonhuman creatures—the *snake*—as central to restoring biological and social equilibrium.[24]

It is likely that Moses' bronze snake spirituality in Numbers—continued by his followers for generations and later repristinated by Jesus—is a survival of the ophidian religion he and Aaron practiced prior to Sinai. Earlier in Exodus, for example, Moses and Aaron are told by God that they will be empowered to perform snake magic when they are challenged by the Egyptian Pharaoh to prove their divine mission. God teaches them how to transmogrify their staffs into snakes. "So Moses and Aaron went to Pharaoh and did as the Lord commanded; Aaron threw down his staff before Pharaoh and his officials, and it became a serpent" (Exodus 7:10). But Aaron's final act of viper shamanism is even more spectacular than his first act: after Pharaoh's court tricksters create their version of reptile magic, Aaron's snake becomes all powerful and proceeds to devour the other snake-staffs of Pharaoh's sorcerers. Shamanistic snake religion is a structural feature of Moses' and Aaron's ministries, proving the sacred value of this particular reptile—as both a bronze representational image and a living creature—to be a medium of God's power and presence in Israel, in both the First and Second Temple periods.

In the light of these examples of biblical snake religion, my point is that the crisis of the golden calf is not that Aaron molded an animal image out of precious metals but that he and the Israelites used that image to erroneously report on the details of the Hebrews' deliverance from Egypt. Aaron's mistake is not that he practiced sacred animal religion but that he misidentified the means of the people's deliverance: unlike the bronze serpent, which *did* deliver the Hebrews from destruction, the molten calf did not perform this restorative function. The problem is that Aaron and the people used the spirit-animal of the calf to shift their focus away from the particular means employed by God on their behalf in their escape from Egypt. If all things, manufactured or living, are filled with

God, arguably, then the golden calf *could* have delivered the people from Pharaoh's evil intentions—just as, two years later, the bronze serpent *did* save the people from deadly snakes—but according to Exodus, God did not use an animal symbol in this case but instead enabled the people's escape through an angel (14:19) and a series of alternative ecological energies, namely, a strong east wind (14:21) and pillars of fire and clouds (14:24). The problem is not the golden calf per se but its role as a displacement of the other means—both angelic and environmental—God in fact did use to make possible the Israelites' deliverance. The problem, in other words, is not the totem image as such but the substitutionary use for which the image was employed.

Just as in the case of the bronze serpent, the Bible has no prohibitions against making things out of gold and silver as objects of special veneration—indeed, even as living embodiments of God in material culture. For example, the ark of the covenant—made out of gold, silver, bronze, and other precious metals and gemstones—houses the sacred tablets and the presence of God, according to Exodus 25. We could call it an "animated sacred object" because the ark is the full embodiment of God—the manufactured but vivified thing that God actually *becomes* amid the covenant people—the living object about which God says, "There [in the ark] I will meet with you. . . . I will deliver to you all of my commands to the Israelites" (Exodus 25:22).

Similarly, insofar as the Bible has no prohibitions against rendering God as a living artifact, it also has no all-encompassing prohibition against making God present in animal or animal-like form. In Exodus 25, the ark features two large winged angels made out of hammered gold, and its companion piece, in Exodus 30, is the altar of incense that is covered in animal horns. Far from disparaging animals such as calves (remembering here the trauma of the golden calf incident), the Bible is suffused with laudatory bovine and taurine imagery—from the many stories about the proverbial fatted calf used in ritual slaughtering in Genesis 15, 1 Samuel 28, and Luke 15 to the sacrificial red heifer or bull ceremony in Numbers 19. In like manner, the lion is also one of many other examples of sacred animal iconography in the Bible. Judah, the fourth son of the patriarch Jacob-Israel and progenitor of King David, is named a young lion by his father in Genesis 49:9; and, analogously, Jesus is named the lion of the tribe of Judah, a messianic title that references his continuation of David's monarchial lineage, in Revelation 5:5. The lion, then, is a symbol of both Hebrew royalty and the Christian Godhead. Into the present, the Lion of Judah is the symbolic coat of arms of modern-day Jerusalem, Rastafarian practice, Ethiopian political monarchy, Freemasonry, and even contemporary Christian tattoo art.

Nodding to sacred snake, calf, bull, and lion imagery, my point is that the Bible does not condemn the veneration of animals or objects, whether found in nature or built by humans. Leading up to the trauma of the golden calf incident at Sinai, we have seen that God manifests Godself biblically as the winged Spirit in creation, the nighttime combatant in Genesis, the burning bush in Exodus, and the healing and devouring serpent, also in Exodus. God, alternately, is avian, human, botanical, and ophidian, as I have said. So, in the light of God's bountiful incarnational behavior, the problem of the golden calf is not the conflation of God and the animal kingdom but Aaron's thoughtless presentation of the calf shrine as the means of Israel's deliverance from Egypt. It is the use of the calf, not the calf itself, that God condemns. It is not, therefore, that God cannot be symbolized by, or actually *be*, this or that particular living being or thing but rather that one should be careful to rightly identify the particular being or thing that functions as a divine medium and, thereby, be valued or worshipped accordingly.

## *Christian Animism*

In the light of Moses' snake shamanism—and, as I have suggested, given the absence of consistent prohibitions against the veneration of sacred animals in scripture—I return to the parallelism between the winged God of Genesis and the Gospels to offer, biblically speaking, that *God is flesh*—and, in the particular texts I am highlighting here, bird flesh. God embodying Godself in creation as a cosmic avian being, on the one hand, and at Jesus' baptism as a nesting dove or pigeon, on the other, contradicts the anthropocentric chauvinism of traditional Christianity. Historically, as we have seen, Christianity evolved into a sky-God religion often opposed to the world, the flesh, animals and plants. But a recovery of the accounts of divine avifauna in the Bible—and especially the Gospels—shows that Christianity is rooted in the carnal reality of God in all things. At its core, Christianity is closer to the spiritual animism of first peoples—the belief that all things, including so-called inanimate objects, are alive with sacred presence—than to the otherworldly bias of some strains of religious life and thought. Could it be, then, that Christianity, ironically, is not a world-denying faith but a fully embodied form of animist religion?

My thesis is that animism is Christian faith's lost treasure. Buried within the deepest strata of Christian beginnings is the founding imagery of God bodying forth Godself within creation's Spirit, Genesis's late-night combatant, Exodus's burning bush, Moses' snake staff, Jesus' flesh, and Luke's baptismal dove. To be sure, later theological traditions distanced themselves from Christianity's formative origins and began to empty the world of signs of corporeal divinity. But the *ad fontes* cry of this book consists of my sustained effort to recover the rich trove of stories about God's embodiment within the natural world as central to both formative and contemporary Earth-loving Christianity as well.

To me, the most compelling religious thought today works the interface between Christianity and animism. An exceptional example of this effort is the ecotheology of the Native American Christian writer George E. "Tink" Tinker. Tinker offers an overall reframing of Christian doctrine using the model of Earth as a living sacrament that gives of itself (or herself) for the welfare of all beings. Using an animist worldview, Tinker infuses Eurocentric Christianity with the lifeblood of Indigenous stories that celebrate all creatures as bearers of divine worth. In particular, Tinker writes about nature as the self-sacrifice of "Corn Mother"—the divine name for Earth's sustaining power and sustenance in the first peoples of eastern and southern North America, including New Mexico—who gives of herself in the nourishing plenitude of plant and animal foods everywhere. In Indian Christian theology, "Corn Mother" is the name some Indians give to the manifestation of Christ in the world's bounty. In this reframing of Christian belief in animist terms, the act of eating is participation in the *interspecies* reality of Christ-as-Corn-Mother's sacrifice—the gift of her earthen body so that all interrelated beings can live and flourish. Tinker writes,

> Out of this story emerges a considerable theology that includes the important teaching that all life—including that life considered un-alive by Euro-science: rocks, rivers, lakes, mountains and the like—is inter-related. When one finally understands this teaching—a simple sounding notion that requires years (if not generations) of learning—one finally understands the sacramental nature of eating. Corn and all food stuffs are our relatives, just as much as those who live in adjacent lodges within our clan-cluster. Thus, eating is sacramental, to use a Euro-theological word, because we are eating our relatives. Not only are we related to corn, beans, and squash, since these things emerge immediately out of the death of Corn Mother, but

> even those other relatives like Buffalo, Deer, Squirrel and Fish ultimately gain their strength and growth because they too eat of the plenty provided by the Mother—eating grasses, leaves, nuts, and algae that also grow out of Mother's bosom.[25]

Tinker uses the Native American experience of First Mother's gift of herself in nature's bounty to deepen the Christian idea of the *sacrament.* Now "the sacrament" is not only a discrete ritual within organized church life—such as Eucharist or baptism—but also a celebration of the deep eat-and-be-eaten relationships that obtain for all beings within the created order—plants, humans, nonhuman animals, and all others. The belief in the sacramental character of all life, the reality of an interfamilial food web, praises the gift of the sacred whenever we eat, and are eaten by, our wider family of biotic and abiotic kinfolk. For Tinker, the biblical God of creation and the agricultural deity of American Indian life are yoked together in the gift economy of healthy predator-prey interactions. Now Christ's flesh is nature's abundance; Corn Mother's body is nature's generosity; and the God of the Bible and Mother Earth are one.

At times, historical ascetic Christianity has emphasized making room for God by denying the world and the flesh. But this emphasis runs aground on the shoals of the biblical animist belief in the living goodness of all inhabitants of sacred Earth. In stark contrast with world-denying religion, Christian animism offers its practitioners a profound vision of God's this-worldly identity. This perspective argues that all things are bearers of divinity insofar as God signaled God's love for creation by incarnating Godself in Jesus and giving the Holy Spirit to indwell everything that exists on the planet. The miracle of Jesus as the living enfleshment of God in all things—a miracle that is alongside the gift of the Spirit to the world since time immemorial—signals the ongoing vitality of God's sustaining presence within the natural order. Thus, God is not a bodiless heavenly being divorced from the material world but manifested as Corn Mother, who gives of herself in the regular sustenance of Earth's goodness; as the avian Spirit, who protectively alights on Jesus at the time of his baptism; and as Jesus himself, the self-sacrificing serpent-God, who shows God's love through his willing death on the cross.

## *Divine Subscendence*

Nevertheless, this vision of a subscendent God challenges traditional understandings of Christian origins as fundamentally opposed to animism. To maintain this opposition, the standard paradigm for understanding Christianity focuses on a faraway time when Christian civilization supposedly vanquished ancient paganism. In antiquity, so goes the argument, every living thing was experienced as alive with its own guardian spirit until Christian missionaries arrived to destroy the indigenous cultures and beliefs of primordial people. The historian Lynn White Jr., arguably the fountainhead of today's emerging academic field of religion and ecology, advanced this myth of Christian beginnings in his seminal article "The Historical Roots of Our Ecologic Crisis." White writes,

> Christianity, in absolute contrast to ancient paganism and Asia's religions (except, perhaps, Zoroastrianism), not only established a dualism of man and nature but also insisted that it is God's will that man exploit nature for his proper ends. At the level of the common people this worked out in an interesting way. In Antiquity every tree, every spring, every stream, every hill had its own genius loci, its guardian spirit. These spirits were accessible to men, but were very unlike men; centaurs, fauns, and mermaids show their ambivalence. Before one cut a tree, mined a mountain, or dammed a brook, it was important to placate the spirit in charge of that particular situation, and to keep it placated. By destroying pagan animism, Christianity made it possible to exploit nature in a mood of indifference to the feelings of natural objects.[26]

White's formulation of the Christian-animist opposition is generally accepted by most scholars of religion today, as we saw in the Introduction in the work of Graham Harvey and Bron Taylor, but it rests on a simple fallacy: an acceptance of the guardians of Christian orthodoxy's self-understanding of biblical faith's triumph over its pagan, animist origins. As the Christian apologist Pat Zukeran writes, "From Genesis to the present the biblical worldview has clashed with the worldview of animism."[27] Similarly, the theologian Robert A. Sirico writes that "the whole of the religious tradition of the West," in its victory over animism, now argues "that nature has no soul and is not in the salvific plan of God."[28]

Similar to these defenders of Christian traditionalism, and akin to religion scholars such as Harvey and Taylor, White's setup of the triumph of Christianity over animism is inattentive to historical complexity and contemporary nuance. To be sure, Western history is replete with images of prominent holy men who cut down groves of sacred trees, among other destructive acts, in order to desacralize nature among pre-Christian peoples. But while the axe-wielding saint is a powerful and disturbing antienvironmental image in Christian history, this trope only tells one side of the story. As the theologian Belden Land writes, we are all familiar with legends telling "once again the defeat of the natural world through the power of the cross."[29] But then Land continues with a different story:

> This all-too-frequently repeated narrative distorts the actual importance of trees, and of nature generally, in the history of Christian spirituality. For every story about saints who cut down trees in an act of anti-pagan triumphalism, there are two stories of saints living in hollow oaks, singing the holy office along with their arboreal friends, even causing the trees to burst into leaf in deep midwinter. If St. Martin of Tours allowed himself to be bound to a stake in the path of a falling sacred pine (though on being cut, of course, it fell in the opposite direction), Saints Gerlach, Bavo, and Vulmar were celebrated for living in hollowed-out trees, St. Victorinus for causing a dead tree to blossom at his death, and St. Hermeland for driving caterpillars from the forest she loved.[30]

Images of the crusading saint with axe in hand make for good newspaper copy, but they say little about the thoughtful efforts by Christians and other people of faith who valorized in the past, and do so today, the Earth as a blessed community of living beings deserving reverence and respect.[31] Thus, in the contest between emerging Christian and Indigenous animist societies, White et al. fail to comprehend the subtly organic and evolutionary nature of the clash between, and at times the marriage of, these two cultural worldviews—an evolution in which differences were often negotiated instead of being frozen into violent polar opposition.

Far from Christianity exercising victory over animism in its efforts to exploit the living world, the actual beginnings of Christian faith are rooted in the deep soil of an animist vision of interconnected, sacred nature. At its core, pace White and many others, Christianity did not *annihilate* animism in its putative dualistic theology but rather *sublated* it to its articulation of the one God now enfleshed within this world through the human person of Jesus (and, as I have suggested, through the animal person of the Spirit). By saying Christianity sublated animism, I do not mean that it subordinated animism to itself but, rather, that it preserved animism's reverence for sacred nature within the horizon of its own incarnational belief system. This sensibility is present, for example, in Paul's sermon in Athens when, quoting the spiritual writers of his day, he preaches that the "God who made the world and everything in it . . . indeed, he is not far from each one of us. For 'In him we live and move and have our being'; as even some of your own poets have said, 'For we too are his offspring'" (Acts 17:24, 27–28). Here Paul critically correlates the Greek philosophical idea that all things are *in* God with the Jewish and Christian story that the one God of heaven and earth *created* all things. The God who is one is also the God within whom the many subsist. In Paul's theology, the God of all creation, according to the biblical witness, is now as well the all-encompassing Earth divinity within which we live and move and have our being, in the manner of pan*en*theistic animist spirituality.[32]

Formative Christianity, therefore, both preserved and transposed animist sensibilities into a new and emerging biblical idiom. This transposition was played against the incarnational baseline of Christian discourse vis-à-vis its contestation and cohabitation with animist cultures. In this regard, nascent Christianity both transformed and was transformed by primordial Earth-based religion. My reference here to Paul's panentheism in Acts 17 and my overall focus on the Holy Spirit as an avian form of divinity make this point in the biblical register. But one can just as easily make this point with reference to other New Testament animist images that were popular in the immediate aftermath of the postbiblical period in Christian history.

## *Avian Spirit Possession*

In this vein, consider "The Martyrdom of Polycarp," a mid-second-century CE avian-spirit-possession story that focuses on a Christian martyr named Polycarp, a prophet and bishop, from Smyrna, Asia Minor (today the city of Izmir in modern Turkey). "The Martyrdom of Polycarp" is an important animist "bridge" text in Christian history that links theologically the close of the New Testament in the first century and the emergence of the writings of the apostolic fathers, such as Justin Martyr and Irenaeus, in the second century. In the account, Polycarp is commanded to worship the gods of the Roman state under orders of the area proconsul. When he refuses, he is lashed to a stake and prepared for a funeral pyre. At this juncture, a great billowing ball of fire is set around Polycarp by his enemies, and then

> he was in the center, not like burning flesh, but like baking bread or like gold and silver being refined in a furnace. And we perceived a particularly sweet aroma, like wafting incense or some other precious perfume. Finally, when the lawless ones saw that his body could not be consumed by the fire, they ordered an executioner to go up and stab him with a dagger. When he did so, *a dove came forth*, along with such

> a quantity of blood that it extinguished the fire, striking the entire crowd with amazement that there could be so much difference between the unbelievers and the elect.[33]

Horribly, the early Christian Polycarp, who refused to bend the knee to the Roman emperor, is put to the stake to be burned for insurrection. But his burning flesh is not consumed and instead produces a sweet fragrance redolent of wafting incense or freshly baked bread. *Next his chest opens wide, and a dove, almost certainly an allusion to the Holy Spirit, flies forth in triumph.* The winged dove emerging from Polycarp's chest is an instance of "spirit possession" in which the body of a religious practitioner is occupied by a god or spirit or some other type of transcendent being.[34] At least temporarily, Polycarp is inhabited by an animal spirit, most likely the avian God of Jesus' baptism, and then out pours such a copious amount of blood that the fire is quenched, routing the emperor's minions (the "lawless ones") and convincing the onlookers of the truth of Polycarp's devotion. "The Martyrdom of Polycarp" is a Christian animist feast. All four of the primary elements of life—earth, air, fire, and water (or blood)—are charged with sacred power. The story is not an instance of otherworldly Christian faith but a paean to creation's sacred elements—purgative fire, precious metals, bread-like flesh, aromatic incense, cleansing blood, and avian Spirit—now fully integrated to show that God, earthen elements, humankind, and animalkind are one.

The anthropologist Leslie Sponsel writes that "with its incredible spatial and temporal range, Animism arguably qualifies as the great, major, or world religion, instead of Buddhism, Christianity, Hinduism, Islam, or Judaism."[35] From this perspective, animism is Christianity's home ground, its encircling family, its native habitat. In its disputation and dialogue with this habitat, Christianity synchronized with, polemicized against, appropriated, co-opted, misunderstood, at times made war against, but, in general, continuously reaffirmed its sense of family belonging to its animist mother religion. Far from overcoming its animist origins, it instead found ways to translate these origins into a profoundly nature-based sensibility that survives all attempts, by White and others, to drive a wedge between the two traditions. White, along with today's defenders of orthodoxy and many religion scholars alike, argue outright or imply that foundational Christianity emerged ex nihilo from the ashes of conquered animism. But early texts such as "The Martyrdom of Polycarp" are predicated not on a war between biblical and natural religion but, rather, on a mutually transformative dialectic between the Christian tropes of faith and martyrdom, on the one hand, and Earth-religion motifs such as fire, flesh, blood, and avian divinity, on the other.

### *Return to the Crum Woods*

Alongside the Introduction, my point in this chapter has been to argue that Christianity is an animist religion that celebrates the enfleshment of God in many forms and, in particular, in avian form. In this regard, my aim is to reawaken an emotionally felt and primordial sense of spiritual belonging to the wider natural world. In turn, my hope is that this deep sense of belonging to Earth, to God's flesh, as it were, will enflame the heart and empower the will to enable commitments to healing and saving Earth—or creation, as Christians and other people of faith understand the natural world. My point is simple: if God is, among other things, avian, vegetal, and reptilian (if God is winged creator, baptismal pigeon, burning bush, and bronze serpent) and if all things God made (birds, shrubs, snakes, and all other beings) are God-in-the-flesh, then it behooves each of us to care for the natural world insofar as this world *is* God in bones, feathers, skin, soil, air, water, leaves, flowers, eggs, and scales.

The moral implications of Christian animism are profound. All life is now deserving of our care and protection insofar as the world is envisioned as alive with sacred animals, plants, and landscapes. From the perspective of nature-based Christian faith, the Earth is the good and holy place that God made and that we then are enjoined to watch over and cherish in like manner. Saving the environment, then, is not a political issue on the left or the right of the ideological spectrum but, rather, an innermost passion shared by all persons of faith and goodwill who live in a world damaged by anthropogenic warming, massive species extinction, and the loss of arable land, potable water, and breathable air. This passion is inviolable. To me, it flows directly from the heart of Jesus' life and teaching that God is a carnal, fleshy reality who is fully and extravagantly incarnated within all things—thereby making the whole world a sacred embodiment of God's presence—and worthy of our affectionate concern.

The evolutionary biologist Stephen Jay Gould writes beautifully the following: "We cannot win this battle to save species and environments without forging an emotional bond between ourselves and nature as well—for we will not fight to save what we do not love (but only appreciate in some abstract sense). . . . We must have visceral contact in order to love. We really must make room for nature in our hearts."[36] Gould is right: the environmental crisis we now face, at its core, is less a scientific or technological problem and more a spiritual problem, because it is human beings' deep ecocidal dispositions toward nature that are the cause of Earth's continued degradation. The crisis is a matter of the heart, not the head: market values have overtaken community values, and the lives most of us lead in the developed world run opposite the crucial insight in the American Indian proverb "The frog does not drink up the pond in which it lives." Regarding the environmental crisis as a spiritual crisis, my hope in this book is to recover the biophilic celebrations of sacred Earth, in the Bible and Christian tradition, that stand as a countertestimony to the utilitarian attitudes toward nature, and toward ourselves as natural beings, that now dominate the global marketplace.

In this regard, consider Rabbi Arthur Waskow's quotation of a poem about sacred Earth from a Hasidic rebbe two hundred years ago. The rebbe was from the town of Chernobyl, the theological center of Hasidism, a mystical movement in early modern Judaism that saw God alive in myriad forms within the natural world. While Jews in Russia, Ukraine, and Poland looked to Chernobyl for Hasidic guidance, the movement in its heartland was eventually stamped out by Christian pogroms, the Russian Revolution in 1917, and subsequent Nazi persecutions. But today, ironically, in spite of the city's Hasidic heritage, Chernobyl has become synonymous with the worst nuclear disaster in history. In 1986, the Chernobyl nuclear power plant exploded, killing dozens of people, leading to thousands of cases of thyroid cancer, and producing ongoing and untold devastation of the surrounding ecosystem. Yet in spite of the explosion, "Chernobyl" remains synonymous with another expression of power—not the power of nuclear energy run amok but the power of the Chernobyl rebbe's vision of a world suffused with divine presence and beauty.

> What is the world?
> The world is God,
> wrapped in robes of God,
> so as to appear
> to be material.
> And who are we?
> We too are God,
> wrapped in robes of God,
> and our task is to
> unwrap the robes and
> dis-cover that we and all the world
> are God.[37]

Today, however, this world that is God, wrapped in the robes of God for all to reverently unwrap and dis-cover, is threatened by apocalyptic changes that challenge the very existence of everything we hold dear in creation, whether we are people of faith or animists or not. Jesus said in his Sermon on the Mount, "Consider the lilies of the field, how they neither toil nor spin, yet I tell you that even Solomon in all of his glory was not clothed as beautifully as these lilies" (Matthew 6:28–29). Jesus saw everyday flowers as signs of God's beauty in wild nature—a more satisfying aesthetic feast for the eyes and heart than any built structure imaginable in ancient times. (In this passage, as we saw earlier, Solomon is a metonym for spectacular royal palaces and the like.) But rampant climate change is creating cascading waves of extinction for wildflowers along with all other plants and animals across the planet. One wonders, could we still see and appreciate today the field lilies Jesus prized when more than two-thirds of the world's uncultivated plant populations are crashing due to climate-change-induced habitat loss, introduced species, and agricultural pesticides?[38]

Global warming—the trapping in Earth's atmosphere of greenhouse gases such as $CO_2$ and methane from car and power-plant emissions—is propelling air and water temperatures to rise catastrophically, as much as three to ten degrees Fahrenheit this century, resulting in the melting of arctic ice and a rise in sea levels, already more than eight inches since the end of the industrial age and now by three feet or more by the end of the twenty-first century. Today, rising temperatures are also causing unpredictable wildfires and megadroughts; terrible flooding in low-level areas from Bangladesh to the Eastern Seaboard of the United States; the world's oceans to become more acidic and thereby lethal to coral reefs and fish stocks, so that at this time, 90 percent of all large fish are gone from the oceans; and, in general, a massive die-off of species similar to the last mass extinction event over sixty-five million years ago, when the great dinosaurs were wiped out. Today's global wipeout of plant and animal populations—what many scientists are calling the "Sixth Great Extinction"—is a biocidal runaway train, with biologists conservatively estimating that thirty thousand plant and animal species are now driven to extinction every year,[39] including, perhaps, the successors to the beautiful field lilies that Jesus lifted up as fluorescent icons of God's benevolent care for all of us.[40]

Stephen Jay Gould is right: we will not save what we do not love. The only hope for our collective commitment to saving Earth from our exploitative habits is to fall in love again with the myriads of creatures and landscapes that populate the living places that we all inhabit.[41] Our only hope for blunting the impacts of climate change is to feel our way back into an emotional relationship with our landed, aquatic, and aerial kinsfolk. This is the religious and ethical promise of Christian animism: a new vision of a world in which all things are steeped in God's presence and thereby deserving of our solicitude or, even better, our love. *We will save what we love.* The first naivete of primordial animism is lost to most of us, but now a second naivete of biblical animism is available as a critical but innocent affective disposition toward nature sufficient for rekindling our spiritual bonds with our plant and animal relations.

Nurturing loving and healthy bonds with the many others starts with me in the natural world—the world charged with divine biopresence. Jacob saw God "face to face" in the visage of his night wrestler. Moses experienced God in natural form through standing on "sacred ground" where he heard God's voice in a fiery bush. And perhaps remembering the God-bird of the Genesis creation, John's followers saw God "in bodily form, as a dove" (Luke 3:22), at Jesus' baptism in the Jordan River. In like manner, I too yearn to see and hear God in my own time and place, and so I spend long periods, many summer days, sitting in a big chair perched at the edge of the Crum Woods longing to catch a glimpse of, or at least to hear, the wood thrush singing its song of intoxicating polyphony. When I gently rock in my chair to the wood thrush's supernal rhythms, I take a break from my mad quest for productivity in so much of what I do, and I soulfully drift into a never-ending sequence of notes that stills my spirit, calms my body, and fills my heart with joy and wonder at the beauty of creation. My hope is to understand the thrush's singleness of purpose and the living of its life as a kind of art form—goals I seek to embody in my own life.

In the Sermon on the Mount, Jesus also speaks about wild birds in addition to wildflowers and says, "Consider the birds of the air, they neither sow nor reap nor gather into barns, yet your heavenly father feeds them" (Matthew 6:26). To rekindle my desire to nurture the sacred Earth, I consider one particular bird cared for by the heavenly father, the wood thrush, often repeating to myself "The Peace of Wild Things," a poem that the farmer-philosopher Wendell Berry wrote concerning the refuge he finds among his own feathered kinsfolk in Kentucky:

> When despair for the world grows in me
> and I wake in the night at the least sound
> in fear of what my life and my children's lives may be,

I go and lie down where the wood drake
rests in his beauty on the water, and the great heron feeds.
I come into the peace of wild things
who do not tax their lives with forethought
of grief. I come into the presence of still water
and I feel above me the day-blind stars
waiting with their light. And for a time
I rest in the grace of the world, and am free.[42]

Like Berry, especially when I am distraught and feeling hopeless about so many things—my family, my students, my work, my country, Earth's future, and much more—I take refuge in the exquisite grace of the wood thrush. I take refuge in this lovely, feral creature whom God feeds in order to remind myself that God seeks to care for all of us, avian as well as human, and that this is the ground of our hope in a depredated but still beautiful world. And I ask myself, if God was *once* the nesting, brooding bird-God of biblical antiquity, present at the dawn of creation and Jesus' baptism, could not God *today* be the ethereal wood thrush who lives in the Crum Woods? In a world on fire—in our time of global warming or, better, global dying—I wager everything on this hope.

CHAPTER 2

# The Delaware River Basin

### *Toxic Tour*

I recently took a group of Swarthmore College students on a "fracking tour" within the Endless Mountains of northeastern Pennsylvania. One of the most awe-inspiring scenic areas in Pennsylvania, the Endless Mountains are a green expanse of old-growth forests, covered bridges, spectacular waterfalls, rocky cliffs, and lush wildlife. Along the way, we visited Ricketts Glen, which is one of my favorite stopovers in the Endless Mountains. The glen (Scottish for "narrow valley") is a remote nature preserve of five-hundred-year-old hemlock and oak trees, brightly colored lichens and mosses, and twenty-two waterfalls that crisscross a raging mountain river with slippery walking stones and winding pathways.

But the preserve and surrounding mountains are also the site of a lucrative experiment in extreme energy extraction: the deployment of hydraulic fracturing (or "fracking") technology to crack open and recover deep-rock deposits of natural gas and oil. Fracking gas and oil fields in the Endless Mountains is especially profitable for developers because this area floats above an underground sea of untapped natural gas reserves. These reserves, called the Marcellus Shale, are embedded in the sedimentary rock that extends throughout the Appalachian Basin from New York to West Virginia, including much of Pennsylvania. As the environmental journalist Tom Wilber writes, "In short, the [Marcellus] rock holds one of the largest gas fields in the world in the middle of one of the largest energy markets, including the metropolitan areas of New York, Philadelphia, Pittsburgh, and Boston."[1] I wanted my students to see firsthand the waterfalls, vibrant bird life, and picturesque back roads of the Endless Mountains against the backdrop of the compressor stations, pipelines, and drilling rigs that now pockmark the landscape, turning a pristine rural community and nature refuge into an industrial fossil-fuel extraction zone of massive proportions.

On the fracking tour, we met a local dairy farmer named Carol French from Bradford County, Pennsylvania. Carol told us about the impacts on her health, family, and livelihood that Chesapeake Energy Corporation created by drilling for natural gas on her property and in her neighbor's fields near her farm. Some years ago, Carol and her neighbor agreed to contracts that consigned the mineral and gas rights under the ground of their homes and farms to Chesapeake Energy. What seemed like a windfall at the time—Carol received $85 per acre for a total of $13,600 for selling her property's mineral rights—turned into a nightmare. To begin with, Carol's well water became contaminated by the gas drilling nearby. During our tour, she brought to show us a jar of gelatinous, milky water that she said came from her house well. Her drinking water for family and farm animals used to be pristine, but now, she said, "Our water turned white. When you let it sit for about three hours it will have sand in it. When it comes out of my faucet it'll have like a sand and a green, mossy carpet and then it will gel at the top. I never had that problem before, never had water that looked like gelatin before."[2] In building gas wells within a two-mile radius of her home, Carol said that Chesapeake Energy's pollution of her ground water led to health problems for her and her family and her dairy cows. She showed our group a terrible red rash that extended along both of her forearms—a rash that now afflicts her after the drill sites were established in her neighbor's open fields. "I have a rash, several rashes—the same ribcage rash that's on my cows too. I don't know what's happening to them."[3]

But as difficult as the stories about the contaminated water and skin lesions were to hear, what I found most disturbing about Carol's narrative was her recounting of her daughter's health problems and her daughter's concomitant decision to move away from the affected area.

> In October [2009] they were drilling a mile and a half away and my water was white, day and night, for a whole week. My daughter got sick. She ended up having sharp, stabbing pains and they'd move all around her stomach. She had a high fever for three days and diarrhea which turned to blood. She lost ten pounds in seven days. On the seventh day she asked me to take her to the hospital and they found a lot of fluid floating in her abdomen. They found her right ovary, her spleen and her liver were enlarged. The doctors said they didn't know what she has.
>
> My daughter left home to look for another one. She found a place in Tennessee and got a job. December 26th, she left me. She left Pennsylvania. While she was looking for another home she realized it was our water, because she started feeling better. Her fever, her diarrhea, all her aches and pains went away.[4]

Did Carol French and her daughter, family, and dairy cows get sick from naturally occurring seeps of swamp gas or other organic compounds that just happened to get released into their drinking water at the same time Chesapeake Energy began drilling on her neighbor's land? Or, as seems more likely, did the Frenches' groundwater become compromised with the advent of the nearby petrochemical well operation, leading to long-term health problems as well as the decimation of Carol's family business and way of life?

Carol told our group that state and local authorities neither are interested in pursuing nor have the legal authority to initiate a thorough investigation into compromised water complaints based on gas drilling. Perhaps she was aware of the Energy Policy Act of 2005, which, among other controversial measures, "exempted the natural gas industry from the Safe Drinking Water Act, which governs what can be injected into the ground."[5] This 2005 act effectively allows the booming gas-shale industry the right to embargo any information about the types and quantities of chemical drilling fluids used to pull gas from, among other places, the Marcellus formation. The provision in the 2005 act that grants the industry an exemption from the Safe Water Act of 1974, and thereby restricts the release of fracking data and its community impacts, is often referred to as the "Halliburton Loophole," named after the giant multinational energy services company. In the early 2000s, then–vice president Dick Cheney, who had previously served as Halliburton's CEO, ran an energy task force in the White House that sandwiched into the 2005 Energy Policy Act the provision that permitted the emerging shale-drilling industry the right to keep silent about the toxic mix it was injecting into potential gas fields.[6]

Does the "Halliburton Loophole" explain why, after repeated inquiries and trips to Pennsylvania's Department of Environmental Protection (DEP) and the Environmental Protection Agency (EPA), Carol could not get any information about the chemical makeup pumped into the shale formation underneath her dairy farm that, most likely, contaminated her groundwater, leading to her health problems and the breakup of her family? Recently, in response to injured parties such as Carol French, the EPA said that while it "found specific instances where one or more [fracking] mechanisms led to impacts on drinking water," nevertheless, it "did not find evidence that these mechanisms have led to widespread, systematic impacts on drinking water resources in the United States."[7] But my fracking tour group's meeting with Carol suggests otherwise, namely, that hydraulic fracturing has generated a sorry pattern of contaminated wells and aquifers that people rely on to meet daily needs. Carol concluded her story to our toxic tour group: "After what happened to my daughter, I went down to the EPA and I was going to talk about my concern with the food since I'm in the dairy business, but other people there set out their black and brown water. And I looked at it thinking, this is my water, this is why my baby left me."[8]

My class's trip to the Endless Mountains and conversation with Carol French underscored for me a powerful truth about today's environmental movement: there is no core distinction between climate change, habitat loss, and social justice. With increasing temperatures, continued drought, and rising sea levels, the planet is under threat from the extreme energy companies; in turn, these companies are converting previously unspoiled landscapes in places such as the Appalachian Basin in Pennsylvania into massive mechanized extraction zones that place at risk people's homes and workplaces. Environmental protection and human well-being are flip sides of the same coin. The gas industry's degradation of water, air, and land in northern Pennsylvania is not only a mortal threat to Earth's climate system and to local populations of flora and fauna but also to the health and future of farming families such as Carol French's. To be sure, hydraulic fracturing is an environmental crisis because it has dumped heat-trapping gases into the atmosphere by converting productive farmland and primeval forest into mining zones scarred by heavy industry. But in the Pennsylvania frack fields, Big Gas is also a human disaster because it has destroyed a rural and agrarian way of life for scores of families and workers who have lost their residential security and vocational livelihood. Global warming—and the industry that drives it—is not an abstraction for the people who live in its wake. At first glance, it may seem that fossil-fuel-driven climate change, on the one hand, and the question of safe living and labor conditions for people at home and in the workforce, on the other, are separate issues. In reality, however, the rapacious business model of the extraction industry has a direct impact not only on global warming, generally speaking, but also on the particular hopes of working families for nontoxic domestic and work environments, good wages, and adequate education and health care for all members of the community.

Our current carbon-intensive energy system, aggravated by the growing reach of the gas-drilling industry, should sharpen our awareness of the basic connection between environmental well-being and social ethics. Abuse of land and water in northern Pennsylvania is inseparably interrelated with the exploitation of nearby human communities that depend on this land and water for their livelihood and welfare. While Marcellus Shale drilling is an economic boon to oil investors and occasional local clients, it poses dangerous risks to clean air and water supplies in the affected areas—as well as for everyone else in the mid-Atlantic area who lives downstream from the extraction zone. As the City of Philadelphia Water Department puts it, "Natural gas extraction from the Marcellus Shale presents a significant economic opportunity for communities and landowners above the deposits; however, this drilling will impose currently unknown costs on Pennsylvania's water supplies."[9] Hydraulic fracturing is now taking place in the northernmost reaches of the Philadelphia watershed—the Delaware River Basin—and it is likely that the millions of tons of hazardous chemicals and millions of gallons of toxic wastewater generated by this process have already leached into the water supply, affecting close to two million area residents. New York State, in order to protect its pristine water supply, has banned all fracking within that portion of the Marcellus Shale formation that extends into the southernmost part of the state. The state of Maryland has done the same. But Pennsylvania has chosen a different course: permitting its state forests and farmlands upstream from Philadelphia and the surrounding region to undergo maximal energy extraction in order to serve the avaricious revenue needs of private corporations.

## *Heidegger's Root Metaphors*

The question of how best to use and preserve healthy ecosystems—with reference to the totality of a bioregion's soil, water, air, plants, animals, and people—is posed with force in the light of the Marcellus Shale industry. How to manage responsibly expanses such as the Marcellus trades on a set of "root metaphors" about the nature of the particular areas in question that form the basis of different stakeholders' core assumptions.[10] So these are the burning questions of our time: How do we understand, in our minds and hearts, the value of land and water, plants and animals, and the atmosphere and the climate alike? What are the founding images and ideals that guide our interactions with the biological world? Do we envision nature as *commodity* or *gift*? As *financial asset* or *life-giving grace*? Are land and water *energy reserves* that stand ready for use and exploitation to meet human needs? Or are natural places *communal gifts* to be nurtured and protected for the welfare of all beings, human and more-than-human alike? Do we think of water and land as *developable capital* for short-term economic gains? Or do we think of landed sites as *nature's bounty*, an enduring bequest that enables the long-term flourishing of the entire planetary community? Is Earth primarily *private real estate* whose value inheres in its designation as an article of trade that can be bought and sold in the marketplace by investors and speculators? Or is Earth a *public offering* of benevolence and well-being, a permanent legacy for the sustenance and joy of all of Earth's inhabitants?

This mode of questioning our root metaphors for the living world stems from the philosopher Martin Heidegger's work on nature and technology. Heidegger writes that we bring two dispositional orientations to the natural order of things—participatory *bringing-forth* or coercive *setting-upon*—that shape our interactions with this order. Heidegger's initial insight is that everything is always in flux; whatever *is* is continually evolving, growing, moving, changing. This process of constant motion and activity Heidegger calls "organic bringing-forth." Heidegger's German term for bringing-forth is the verb *hervorbringen*, which he uses to translate the Greek term *poiesis*, which means "to make" and is the root of our English word for poetry.[11] His idea is that all things are in different stages of transmogrification and full becoming; all things are being made or formed in biological emergence over time. Life itself has its own transformational rhythm or kinetic artistry, as it were. For example, a milkweed plant pushing forth into full blossom or a Monarch butterfly metamorphosing into a winged insect, left to their own devices, are ever-changing examples of the process of bringing-forth. These types of creative events are a kind of everyday poetry—or *poiesis*, to use Heidegger's Greek-influenced terminology. Such occurrences happen naturally all of the time and everywhere around us.

Glossing the Greek term *physis*, which is the basis of our English term "physics," Heidegger refers to this type of bringing-forth simply as "the growing things of nature," as opposed to "whatever is completed through the crafts and the arts."[12] Within the overarching dynamic process of *poiesis* or bringing-forth, Heidegger then makes a distinction between organic change (*physis*), on the one hand, and human-mediated transformational processes, on the other. *Techne* is the term he assigns to this second mode of change; it is the Greek word for the creative activity of the artist or craftsperson, and it is the basis of our English notion of technology. It may seem odd to us to regard the idea of technology as having its conceptual origins in ancient Greek practices of the arts and crafts. But Heidegger's point is that all forms of "making," whether natural or human directed, are, as original activities, exercises in poetic creativity (*poiesis*). Nothing, in other words, simply *is* because whatever *is* is always in the process of changing or being changed into something else—whether that process is naturally occurring (*physis*) or catalyzed by human creativity and ingenuity (*techne*).

The problem for Heidegger, therefore, is not that things change according to human technological intervention (*techne*) as opposed to natural evolution and adaptation (*physis*). Technology, understood originally as poetic or creative making or craft, is not a problem in and of itself. The problem is the *attitudinal disposition* we bring to the process of anthropogenic change. Is our attitude one of attunement to organic rhythms so that our involvement in the cycle of bringing-forth is sensitive and respectful of this cycle? Is our practice of *techne*—or technology—thoughtfully attendant to the emerging patterns of relationship that naturally reoccur in the process of *poiesis* or bringing-forth? Or is our technological intervention into nature's emerging process a violent "setting-upon" that process, as Heidegger says, according to our own ego-centered designs—a setting-upon that pays little if any attention to the destructive impacts generated by our intrusions into the more-than-human world?[13]

Heidegger's centering examples of appropriately innovative *techne* are the making of a silver chalice, on the one hand, and the construction of a wooden bridge, on the other. In the first instance, chalices, since time immemorial, have been used as sacred, decorative instruments to facilitate a community's relationship with its God or gods. Fashioned by an artisan out of silver ore, a chalice, rightly produced, is both deeply resonant with the energies of the Earth and beautifully expressive of the silversmith's artistic and religious ideals. For Heidegger, silver ore, as *physis*, can simply remain embedded within a particular geological formation and evolve naturally over time without human intervention. Or the ore can be mined, in the spirit of *techne*, and its potential as a liturgical vessel for mediating the relationship between Earth, human beings, other beings, and God or the gods can be realized through restrained but visionary human agency.

In this manner, the silver ore, through careful human intercession, is enabled to perform a central role, as chalice, within "the sacrificial rite in relation to which the chalice is determined as to its form and matter."[14] The silver *matter* of the chalice and the ornamental *form* it takes is *determined*, according to Heidegger, by the religious *rite* that draws up out of the ground the potentiality of this sacred vessel. The artisan, then, stands in perceptive openness to the possibilities of the silver ore and thereby attunes him- or herself to the shape, function, and end use of the ore in the performance of a religious ritual. In this model, it is as if the silver ore "speaks" to the practitioner of *techne* and thereby mysteriously signifies its "readiness" to be creatively harnessed in the service of spiritual values.

Heidegger's model of attentive *techne* is analogous to Michelangelo's famous sculptures of four slaves whose emerging forms are trapped, it seems, within the material world. Now housed within the Accademia Gallery in Florence, Italy, Michelangelo's four prisoners or slaves sculptures—*The Awakening Slave*, *The Young Slave*, *The Bearded Slave*, and *The Atlas (or Bound)*—depict different figures struggling to be set free from within the unfinished slabs of marble that house them. Each writhing male form is only partially carved out of its rocky home; each shape appears to be yearning to burst forth from its stony origins. Whether these giant pieces were unintentionally left unfinished or, as many critics think, deliberately left incomplete by design, these partially finished marvels of strength and will only gradually appear within the confines of the marble blocks. Michelangelo, according to one of his biographers, envisioned his task as bringing into presence artistic *forms* that were already contained in germ within *formless* matter. In the emergence of Michelangelo's monumental sculptures, it was the rock *itself* that gave rise to the twisted figures within each slab of marble. Michelangelo, then, was only the midwife, as it were, who birthed into existence the twisted shapes already nascently present within the blocks of stone.[15] Similarly, Heidegger writes that the chalice owes its existence as much to the alluring mystery of the silver ore itself as it does to the silversmith who realizes the ore's potential: "Silver is that out of which the silver chalice is made. As this matter, it is co-responsible for the chalice. The chalice is indebted to, i.e., owes thanks to, the silver for that out of which it consists."[16] For Heidegger, the craftsperson only excavates the ore's buried possibilities—possibilities that are indebted to the nature and quality of the ore itself. Together, the silversmith and the silver ore *cocreate* the chalice. Like Michelangelo, who did not create de novo each of the prisoner-slaves but rather discovered what was always-already latently existent in his large chunks of marble, Heidegger regards the artist/technologist as a practitioner of discernment who realizes what *might* be—in Michelangelo's case, the four prisoners or slaves; in Heidegger's case, the silver chalice—by cultivating a patient sympathy for what *is* incipiently present.

Heidegger's other example of creative *techne* is the building of a wooden bridge in the countryside. In general, he writes, large-scale energy projects, such as hydroelectric dams, that radically alter the course of rivers, riparian buffer zones, and nearby fields and towns are instances of exploitative setting-upon, as he calls it, not *techne* in harmony with nature's own bringing-forth. In the case of the Rhine River, a site-suitable bridge made from local materials, and one that connects two shorelines together in a graceful arch, is an instance of artistically appropriate technology; whereas a hydroelectric plant that transforms the river into a monetized reserve of raw power is not. He writes,

> The hydroelectric plant is not built into the Rhine River as was the old wooden bridge that joined bank with bank for hundreds of years. Rather the river is dammed up into the power plant. What the river is now, namely, a water power supplier, derives from out of the essence of the power station. In order that we may even remotely consider the monstrousness that reigns here, let us ponder for a moment the contrast that speaks out of the two titles, "The Rhine" as dammed up into the *power* works, and "The Rhine" as uttered out of the *art* work, in Hölderlin's hymn by that name.[17]

The Rhine carries a double valance of meanings depending on the root metaphors and corresponding technology that is deployed within its watershed. "The Rhine" as power plant, versus the eponymously named poem about the "freeborn river" by Friedrich Hölderlin, makes this point clear. Is the Rhine primarily viewed as a booked asset of raw energy to fuel the industrial state? Or is it seen as a beloved waterway wherein limited technological development actually enhances, rather than degrades, its historical role as a site of beauty, wonder, and power?

As with the silver chalice, Heidegger writes that the natural-materials bridge powerfully "gathers to itself" not only natural elements such as earth and sky in thoughtful and felicitous combinations but also humans, whom he calls "mortals," and God or the gods, whom he calls "divinities." He writes, "The bridge *gathers* to itself in *its own way* earth and sky, divinities and mortals."[18] For Heidegger, sensitively designed things "gather" together what he calls the "fourfold": earth, sky, mortals, and divinities. Mindful, well-built structures and landscapes—indeed, in principle, any carefully imagined form of technology—are not simply buildings for human occupation but places of "dwelling" that nurture sustaining relations between the natural, the human, and the divine order of things.[19] Beautiful, functional spaces and products are "gatherings" for profound relationships between human beings and all other beings within the natural world. Earth and sky, water and sunlight, and people, plants, animals, and spiritual beings themselves—the "fourfold," as Heidegger puts it—are brought together in novel patterns of mutuality and belonging through site-sensitive creativity and technology.

Heidegger's case studies of the silver chalice and the wooden bridge are consistent with what today is called *biomimicry*. Biomimicry is a techno-philosophical approach to solving problems in land use, energy production, and biomedical research through policies and products designed to simulate the natural world's own systems and patterns of interaction. As Norman Wirzba writes, "We can learn, for instance, from spiders who make fibers that are much stronger than anything we produce, yet do it without heat or waste. We can learn from abalones that make polymers, in cool water and without toxins, that are much tougher than any ceramic we currently make. Called 'biomimicry,' this approach learns from, works with, and respects the integrity of creation. In a way, it welcomes the creation to be itself and encourages humans to enter more fully into its beauty and grace."[20] Biomimicry imaginatively borrows models and ideas from nature to solve complex social, environmental, and human health problems. Like Heidegger's chalice or bridge, products and designs that bring-forth, rather than set-upon, nature's benevolent potential are creatively developed. Further examples include siting wind farms in relation to carefully studied land formations and air currents in order to generate appropriate levels of energy; creating so-called permaculture societies in which nature-based design and self-regulating habitat management create environmentally sustainable communities; and developing medicines that emulate the natural resilience of particular plants for addressing human health needs. These biomimicry examples, resonant with Heidegger's own nature-sensitive case studies, demonstrate how technological innovations that are attuned to natural energies can be practically functional, culturally appropriate, and aesthetically pleasing.

## *Calling Spirit from the Deep*

Today, is hope in the world as the gathering of what Heidegger calls the fourfold still available for some of us, even if, for others of us, the time has passed when we can encounter nature as a spirit-filled gathering of the fourfold? While previous generations saw the natural world as a communion of subjects enlivened by indwelling spirit, many of us today see the same world as a jumble of inanimate objects emptied of any traces of spiritual presence. At this point in my argument, let me be clear that there is no way to be certain whether the "communion of subjects" worldview is true and the "inanimate objects" perspective is false. Outside of one's basic assumptions, one cannot prove that the animist vision is right and the materialist viewpoint is wrong.[21] Beyond the horizon of one's founding beliefs, one cannot demonstrate with any certainty to a skeptic that the spirit-filled-world position is superior to the antianimist worldview. Rather, as an "orientation" toward, and not as a "proof" about, the world, I regard animism as an *incantatory gesture* toward the natural order: by opening myself to the possibility of animating spirit within all things, I subsist in the fragile hope that I can summon the presence of numinous realities within the everyday. Transfixed, for instance, by the lilting magic of the wood thrush's song, am I not encountering the wonder of spirit in my midst? Refreshed and renewed in the spray of a towering waterfall in Ricketts Glen, am I not transported into a sacred place inhabited by joyous powers greater than myself? Can I not, therefore, live in hope that the more-than-human world is not dead and unfeeling but alive with intimations of spirit within all things?

Immanuel Kant is famous for asking three questions: What can I know? What must I do? and What may I hope?[22] Each question forms an interconnected framework for living one's life with confidence, integrity, and meaning. In life's three realms, first, I can *know* certain truths about the material world to be certain; second, in the realm of ethics, I learn that I *ought* to perform some acts and avoid others in my everyday relations with the others around me; and third, I discover that I am entitled to live in *hope* that my beliefs about God and the world are true in the religious sphere. While animism makes claims about the material world and our daily ethical obligations therein—namely, that the world is a gathering of ensouled beings who are deserving of our care and reverence—such claims are not empirically obvious or morally certain to the person not inclined to embracing animism. In a Kantian framework, then, animism is sustained by another mode of understanding that we might call *subjunctive* knowing; it is governed by an epistemic appreciation of what *might* be, the wisdom of a kind of *possibility* thinking about the world. Animism, therefore, is buoyed by an intuitive form of comprehending the world that one might call *ecstatic preternaturalism*: the unseen but everywhere-felt resonance with the joyous presence of spirit in commonplace nature. In and with and alongside daily life are alluring intimations of spiritual powers that make demands on my processes of cognition and my ethical responsibilities. I may not be able to prove with certainty that such powers exist (What can I know?) or say exactly how as a moral decision maker I am to comport myself to them (What must I do?), but I can go forward with confidence that my existence will be balanced and purposeful if I wager my life on the belief that these powers should be my daily guide (What may I hope?). Thus, it is in this particular range of my knowing and doing—it is in the registry of hope—that animism finds its natural home.

In this expectant attitude of openness to natural wonder, the practical benefits of invoking the hidden sacred within the created order are extraordinary. Over time, the world loses its character as soulless matter indifferent to the possibility of spirit in its midst. Instead, by calling spirit from the deep of ordinary life, all of reality slowly evolves into a hallowed grove of sentient animals, cherished plants and trees, and much-revered landscapes, mountains, rivers, and seas. Now the world is not a passive, inert object lying ready to be subjected to human domination and exploitation but is rather a breathing, feeling, conscious life source of infinite complexity and beauty that speaks to us in myriad ways about how to live sustainably in and alongside this bounteous gift.

Our graced existence made possible through nature's abundance can follow two paths. One path is hell-bent on exploiting nature's gifts no matter what the costs. Shale gas and oil extraction in the Marcellus basin is a full-scale assault on the fourfold: earth and sky, mortals and divinities. God's landed presence in the Endless Mountains—a stroll across a covered bridge, the taste of fresh cream atop pastured cows' milk, the electric flash of a goldfinch knifing through ancient forest—is under siege by us mortals who have converted many of these sacred mountains and surrounding farms into toxic killing fields. In turn, earth and sky are compromised: earth's soil and water become saturated with dangerous contaminants, and the sky grows dark and bloody with belching smoke and red flares as heavy metals and other poisons are spewed into the vaulted atmosphere that holds and protects all of us. As a scarred and polluted landscape under the control of multinational gas firms, the Marcellus is losing its role as a place that "gathers," in Heidegger's vocabulary, the fourfold realities of earth, sky, mortals, and divinities.

## *Sacrament of Dirt and Spit*

But there is another path available to us. This is the good road that Jesus the shaman walked. This is the road that celebrates the donative wonder of ordinary life—what Heidegger calls everyday bringing-forth as *physis*—wherein all creatures are cared for on a regular basis ("Consider the birds of the air, they neither sow nor reap nor gather into barns, and yet your heavenly father feeds them"; Matthew 6:26). This reality of creation as reliable bringing-forth is complemented by the human capacity, to use Heidegger's language, to benevolently and creatively intervene in organic *physis* through appropriate forms of *techne*. In Chapter 3, I offer a comprehensive study of Jesus' relationship to sacred land and animals. In what follows, I will lay out the case for Jesus as an earth-based healer, in one particular biblical account, through the arts of restorative *techne*. The teaching that the birds are cared for daily by their father vibrates with other biblical teachings and stories that show Jesus to be an astute artisan of God-given matter for the renewal and healing of others.

Jesus' skill and sensitivity in *techne*—his ability to bring-forth nature's inherent potential—is paradigmatically illustrated in the Gospel of John's story of Jesus' healing of the blind man with a poultice he makes out of the soil with his own spittle. Living in the lap of creation's many gifts, Jesus harnesses the curative powers of the earth to do something radical and unexpected within his community. Here are the highlights of the narrative:

> As he passed by, he saw a man blind from his birth. And his disciples asked him, "Rabbi, who sinned, this man or his parents, that he was born blind?" Jesus answered, "It was not that this man sinned, or his parents, but that the works of God might be made manifest in him." . . . As he said this, he spat on the ground and made clay of the spittle and anointed the man's eyes with the clay, saying to him, "Go, wash in the pool of Siloam" (which means Sent). So he went and washed and came back seeing. The neighbors and those who had seen him before as a beggar, said, "Is not this the man who used to sit and beg?" Some said, "It is he"; others said, "No, but he is like him." He said, "I am the man." They said to him, "Then how were your eyes

> opened?" He answered, "The man called Jesus made clay and anointed my eyes and said to me, 'Go to Siloam and wash'; so I went and washed and received my sight." (John 9:1–3, 6–11)

What arrests my attention in this story is Jesus' sacred mud-pie ritual. In Chapter 1, I discussed Moses' role as a shaman who employed snakes in Pharaoh's court and the bronze serpent in the Sinai wilderness to liberate and heal the ancient Hebrews. I noted how Jesus then appropriated Moses' ophidian symbolism to identify himself as the new bronze serpent who offers eternal life to his followers. Similarly in John's story, Jesus continues his shamanistic activity, but now not through serpent imagery but through folk medicine. Just as Moses channeled God's power through snake handling, Jesus actualizes God's power by mixing his saliva with the soil near the pool of Siloam, on the outskirts of Jerusalem, to form a clay-like compound that he then pastes onto the eyes of the blind man. Analogous to Heidegger's silver ore or Michelangelo's marble block, the Jerusalem soil is ready-made for Jesus to realize its inherent medicinal power to heal the blind man. Here Jesus' *techne* skills come to the fore. Spitting on the ground, Jesus creates a sticky, wet salve that he applies to the man's face and then tells the blind man to wash his face in the Siloam pool, wherein the man experiences full sightedness for the first time in his life.

A quickening source of curative renewal and refreshment, the gift of the sacred *ground* we inhabit is essential to all persons' well-being. Not only the ground but our own vital bodily fluids (including *saliva*) are necessary for physical and spiritual health. Nevertheless, many people and societies today regard spittle as a dirty substance and spitting as a nasty habit (it is illegal to spit publicly in Singapore, for example). This is odd, because in the twenty-first century, we fancy ourselves generally comfortable with our bodies and their natural products and secretions. After Darwin, Freud, modern biology, and mass sex education, we think of ourselves as scientifically self-aware and accepting of our core bodily functions. As well, we say that we regard the many emissions that flow out of our everyday physical activities—blood, urine, feces, sweat, semen, vaginal fluids, tears, and yes, saliva—to be the natural fluids of embodied, organic beings such as ourselves. All of this is normal and healthy. Or is it?

It seems, rather, that while we *say* we embrace our bodily excretions, our actual vocabulary for the body *betrays* us. Take the example of so-called sanitary napkins. Menstrual blood is a potent symbol of women's progenerative capacity—a marker of reproductive health and well-being. But modern societies have denigrated menstruation as a dirty, unclean activity that needs to be "sanitized." We claim to be enlightened about normal, healthy bodily functions, but our everyday language about hygiene and cleanliness belies such claims. The word *sanitary* is from the Latin *sanitas*, which means "health." Etymologically, terms today that have their roots in *sanitas* stand for making something healthy or clean. These terms include, for example, *to sanitize*, the sterilization of an object free from filth and contamination; *sanitation*, the proper disposal of sewage and solid waste; and *sanatorium*, a hospital setting for the treatment of chronic disease. Filth, contamination, sewage, waste, disease—sanitary napkins function to "sanitize" female bodies, to protect girls and women from the "unhygienic" bleeding of their own bodies, sending the message that this outward sign of life itself is unclean and unsanitary.

Mary Douglas writes that all cultures, including our own, are comfortable with the body when the body stays intact, its orifices and margins are secure, and nothing is seeping from the body or traversing the boundaries between the body and the outside world. We regard bodies as healthy and clean when they maintain their integrity and do not leak or discharge fluids, but when bodies break down and ooze at their margins, they are considered unhealthy and dirty. Douglas writes, "All margins are dangerous. If they are pulled this way or that the shape of fundamental experience is altered. Any structure of ideas is vulnerable at its margins. We should expect the orifices of the body to symbolize its especially vulnerable points. Matter issuing from them is marginal stuff of the most obvious kind. Spittle, blood, milk, urine, feces or tears by simply issuing forth have traversed the boundary of the body. So also have bodily pairings, skin, nail, hair clippings and sweat. The mistake is to treat bodily margins in isolation from all other margins."[23]

Bodily margins and openings are dangerous, Douglas argues, because they stand for a body out of control, a body that is pockmarked with orifices small and large that at any moment can crack open and threaten not only bodily integrity but the order of society as well. "Bodily margins," she writes, should not be seen "in isolation from all other margins,"[24] meaning that the body, like society as a whole, is healthy when intact and dangerous when its margins break down. Well-behaved, structurally whole bodies are a promise of physical and cultural cohesion as well. The breakdown of intact bodies is a dangerous sign of destabilizing forces outside of one's control that threaten both one's personal well-being and the sense of unity and completeness that social groups rely on for ongoing stability. For this reason, Douglas writes, different societies have different rules for what constitutes dangerous or, in religious terms, polluting behavior. Anything that threatens the integrity of the body is experienced as hazardous to self and others, but social groups differ as to the particular events that cause social pollution and chaos. Douglas continues, "In some, menstrual pollution is feared as a lethal danger; in others not at all. In some, death pollution is a daily preoccupation; in others not at all. In some, excreta is dangerous, in others it is only a joke. In India cooked food and saliva are pollution-prone, but Bushmen collect melon seeds from their mouths for later roasting and eating."[25]

It is not clear, then, why particular activities—menstrual blood in one case or saliva in another—represent occasions for polluting behavior. Julia Kristeva, commenting on Douglas's work, writes that there is nothing inherently dangerous or polluting about a particular boundary-violating discharge or event. She writes, "Taking a closer look at defilement, as Mary Douglas has done, one ascertains the following. In the first place, filth is not a quality in itself, but it applies only to what relates to a *boundary* and, more particularly, represents the object jettisoned out of that boundary, its other side, a margin. . . . The potency of pollution is therefore not an inherent one; it is proportional to the potency of the prohibition that founds it."[26] Instances, then, of chaos or defilement only have meaning as they relate to other elements within the symbolic system of "purity" and "filth" in a particular culture. Filth or pollution or defilement are what they are because of how these events relate to other features of a social system that are regarded as symbolically clean or pure.

Now returning to the topic of Jesus' spittle, we can see how likely it was that his mud-pie behavior was offensive to his society—and how offensive it is in our own society as well. As Jesus was a shaman, some people in his time likely saw his healing ritual as genuinely therapeutic. But others likely regarded it with disgust. Jesus took two of the primary defiling elements within his and our own symbolic order—namely, spit and dirt—and mixed them together to form, at least for some people, a "polluting" or "unsanitary" potion to heal the blind man in John's story. In terms of the Jewish purity code of his time, it is reasonable to assume that Jesus performed a formal act of ritual defilement, but this is not my particular focus here.[27] While Jesus likely violated Jewish first-century cleanliness standards by mixing spit and dirt, my primary interest is on the universal (both ancient and contemporary) experience of repugnance, even nausea, that Jesus' action likely produced—and continues to produce. If Douglas and Kristeva are correct that bodily discharges are very often experienced as disgusting and degrading, then Jesus' spitting into the dirt most likely functioned then, and continues to function now, to shock and disrupt his audience (as well as being a transgression of biblical purity standards). Kristeva writes that most bodily discharges are instances of what she calls "abjection." Abjection, she says, is "loathing an item of food, a piece of filth, waste, or dung. The spasms and vomiting that protect me. The repugnance, the retching that thrusts me to the side and turns me away from defilement, sewage, muck. . . . I experience a gagging sensation and, still farther down, spasms in the stomach, the belly; and all the organs shrivel up the body, provoke tears and bile, increase heartbeat, cause forehead and hands to perspire."[28]

Is Jesus' "disgusting" miracle, therefore, a gagging instance of abjection? Could expectorating onto the ground so much so that he could form a paste from the surplus of his spittle cause his audience to wretch? Does it cause us to gag today? Did he bend down and spew a sizeable amount of foamy, phlegmy spit into the dirt to create his strange medicine? Was it merely spit he spewed onto the ground, or did he cough up something deeper, something more vile and sinister, something more bilious and nauseating, something more like *sputum* than spit, with the aim of making a thick, wet plaster that he could then mold over the man's unseeing eyes? Is Jesus' healing action, ironically, an exercise in abjection?

We know Jesus often used more conventional means to perform healings. In Matthew 8, for example, he simply touches a man with leprosy, and the man is healed; later, in the same chapter, he heals a centurion's servant by direct word without any physical touching or intervention. *But in John's story, Jesus seems to want to drive home the point that the abject is holy.* He seems to want to say, "You regard the soil that nourishes you as filthy and your own body with disgust. But the ground is your life, and you should never experience your body's normal functions and emissions with anxiety and loathing. I want to show you that what you fear has been made by God and declared by God to be wholesome and blessed. The earth is not muck and mire to be despised, and nothing that your body secretes or emits is filthy or degrading. Indeed, the dirt that I walk on, just like the dirt under your feet, and the spit from my mouth, just like your own spit—all such things are sacred and beautiful. My message to you is to love the earth and all of its grimy, soiled inhabitants as God loves you and to love your body and all that flows out of your body in the same manner, because all that God has made, no matter how low and dirty and loathsome and unclean it may seem to you, is holy and good and worthy of your highest esteem and most heartfelt enjoyment—including, and especially, dirt and spit."

In this regard, let us not miss a final clue to the inestimably sacred value Jesus assigns to embodied life—including everyday soil and bodily emissions. Note the important word choices John uses to describe Jesus' action in applying to the man's eyes the wet earthen dressing he makes from saliva and clay. Here the animist theme of sacred earth—or, to be more precise, sacred earth mixed with holy spit—emerges as the story's leitmotif. In verse 6, Jesus "spat [*ptuo* in Greek, which may be an onomatopoeic use of a word sounding like its referent, perhaps the forerunner of *ptooey*, the English slang for spitting] on the ground and made clay [*pelos*, which means "clay" or "mud"] out of the spittle, and put on [*epithesis*] the man's eyes the clay." Similarly in verse 11 we read, "He replied, 'The man they call Jesus made some mud and anointed [*epichrio*] my eyes. He told me to go to Siloam and wash. So I went and washed, and now I can see.'" What is sometimes missed in the English translations of this passage—but is obvious in the Greek of the original text—is the *liturgical force* of the verbs used to describe Jesus' smearing the man's eyes with saliva and dirt. Both of the verbs, *epithesis* (verse 6) and *epichrio* (verse 11), that are used to describe how Jesus applied the muddy salve to the man's face denote a sacred ceremonial activity such as the laying on of hands in an ordination ritual or the rubbing on of oil in a baptismal or healing service. The verbs *epithesis* and *epichrio* can mean simply to put on or to apply something, but throughout the New Testament, and in other ancient religious literature as well, the verbs more often stand for the *sacred act of anointing somebody with transformative power* by means of one's hands and/or through the application of natural elements such as water or oil.

The point of the passage is that Jesus' spit—now mixed into the soil to form a mucky paste—is the bodily discharge he uses to *anoint* the man's eyes and heal the man of his blindness. Like wine and bread in the Christian Eucharist, here saliva and earth are the *holy elements* Jesus uses to channel God's grace to his followers. Like the eucharistic blood and body, here spit and dirt are the *sacramental means* of God's power and love. It is for this reason that Jesus says the man's blindness is not a result of sin but the means by which God's work will be manifested (verse 3).

Jesus is an animist shaman who expectorates in the good soil of creation to make a healing plaster and to teach that all of life—even what we reject as dirty and unclean—is charged with divine wonder and power. Did not the creator God do something similar by going into the primeval dirt, perhaps spitting into it as well, in order to form a clay that could then be molded into the first human being, and breathing into this prototypical mud-man—Adam—the breath of life (Genesis 2:7)? Do not Christians still today signify the power of dirt by receiving ashes on their foreheads at the beginning of Lent, eating the fruits of the soil by taking bread and wine at the Eucharist, and burying their dead in the ground, intoning the dirge of the ancient mud-man, "we commit our beloved to the ground, earth to earth, ashes to ashes, dust to dust," an echo of God's statement to Adam, "By the sweat of your face you shall eat bread until you return to the ground, for out of it you were taken; you are dust, and to dust you shall return" (Genesis 3:20)?

Earth is spiritually charged—and spit is as well. In Jesus' mud-pie miracle, spit is not mere spit but the healing elixir Jesus uses to demonstrate divine power. Likewise, dirt is not simply dirt but the life-giving soil Jesus further enlivens with his spit for healing and renewal. What we reject as dirty and unhygienic—as polluting and abject—God loves and persuades us to love. The story of the blind man in John 9 teaches us, most significantly, to love our bodies and others' bodies—and all of the sticky, colorful, and aromatic fluids and secretions therein—as God's gifts for the joy and healing of all beings.

## *Girard's Fear of Monstrous Couplings*

In Chapter 1, I made the case for God as a nonhuman life-form—a winged dove—based on the depictions of the Holy Spirit in all five Gospel accounts of Jesus' baptism. I contended that it is not foreign to biblical traditions that God is a sacred animal—indeed, that every earthen being is a bearer of divine wonder. My goal has been to demonstrate that all things are alive and sacred and suffused with God's presence and power and to say that this attitudinal disposition toward Earth as holy ground is the basis for healing the suffering among ourselves and the wider community of living beings.[29] I have sought to show that it is organic to Christianity that everyday life is saturated with divine presence—including, and especially, the abject elements within this common life, such as dirt and spit. In this chapter, I have analyzed the role Jesus' mud-pie poultice played in the blind man's healing against the backdrop of Heidegger's model of bringing-forth Earth's potential. The flying bird-God, the spitting savior, the healing clay—these opening overtures are intended to root animist Christianity in the rich soil of the biblical witness in order to signal that the whole of creation, alive and sacred, is thereby worthy of our love and protection.

But unfortunately, as I have said, historical Christianity has often missed the point that God is on the wing and afoot within the natural order, and religious faith has devolved into an otherworldly mind-set that denigrates this world as devoid of God's presence. By carefully separating the "heavenly" realm from "earthly" life, Christians and other people of faith have sometimes sought to divorce what is invisible and spiritual from what is visible and material. And, as I have suggested, this division has come at a price: for many, the world has become a dead place with few intimations of the God-given beauty that graces ordinary existence.

In Paul's sermon in Athens, as I noted in Chapter 1, Paul, according to the author of Acts, says that God "is not far from each one of us, for 'in him we live and move and have our being'" (Acts 17:27–28). But what is the medium for God's presence to which Paul refers here, if not the joy and splendor of divinely infused creation itself? Is it not that which is closest to each one of us: the sun's heat on our backs, the water from our taps, the dirt under our feet, the bracing crystallization of one's breath on a winter morning? Is it not impossible to imagine that this might have been what Paul meant by saying that God is intimate with each of us, the one within whom we live and move and have our being? Are not sun, water, earth, and air the animating elements within which we all live and move even as our very being is daily sustained by these providential gifts? God is near to us, God is not distant from us, God is not split apart from daily life. Indeed, God is closer to us than we are to ourselves. Viscous, tangible, sensuous, even consumable—God is the pulsing, driving life force within the wide expanse of the green world that brings all things together for their, and our, common sustenance.

This animist vision of God and nature, however, is a bridge too far for many religious thinkers who seek to isolate God from the Earth. A primary example of this effort is the contemporary cultural theorist René Girard. I regard Girard as one of the most important antianimist apologists for Christianity in our time. Girard, like Heidegger, provides a powerfully overarching vision of reality, but unlike Heidegger's focus on nature as the gathering point for the fourfold, Girard's focus falls on nature as a site of violence and chaos over and against the historical emergence of the Christian religion in the life and teachings of Jesus. Girard's view of reality begins with a tripartite model of religious history, culminating in the revelation of Jesus as the nonviolent in-breaking of God in the world. Following an arc from G. W. F. Hegel and the historian Arnold Toynbee, Girard sees universal history progressing from (1) the Greco-Roman era to (2) biblical times and now (3) culminating in the realization of Christian civilization in the modern age.

Girard posits the first stage in this evolutionary history as the age of so-called primitive religion, when the gods lash out with fury against their human subjects; the second stage is the time of the Hebrew Bible, when religious violence is still active but less common, especially in the period of the Judean prophets such as Isaiah and Jeremiah; and the third stage is the time of the New Testament, when Jesus repudiates the violent deity of the pagans and, to a lesser degree, the Hebrews, in favor of a new religious vision of forgiveness and compassion toward others. According to Girard's supersessionist unfolding of universal history, the "violence characteristic of primitive deities" is improved on by the Old Testament God, who is still not "entirely foreign to violence," and both of these moments are subsequently superseded by the nonviolent God of the Gospel texts—texts that finally "manage to achieve what the Old Testament leaves incomplete."[30]

While Girard understands Judaism as a mixed religion that both cautions against and endorses divine violence, ostensible primitive religions are violent to the core, and such religions portray the gods as cruel and arbitrary nature divinities who revel in the deaths of their human victims. These pagan nature divinities are often portrayed as half-human and half-animal creatures such as the centaur (man/horse), the harpy (woman/bird), and the satyr (man/goat). For Girard, Greek and Roman religion is an exercise in abject savagery. In paganism, evil, hideous semidivine beings, captive to human-like emotions such as vengeance and bloodlust, stand in absolute contrast with the One True God of Christianity who exists outside the world of animals and nature, and the vagaries of human emotions, in every respect. My recovery of Christianity as an animist religion—in which God incarnates Godself in the flesh of a winged animal (Chapter 1) and in which natural elements, such as dirt and spit, are reimagined as healing sacraments (Chapter 2)—would likely strike Girard as a return to the violent world of primitive nature and the animal gods. The transgressive hybridity that defines Christian animism—the identity-fusion among God, Earth, humans, and animals—flies in the face of Girard's indictment of the ancient religions as sinkholes of monstrously grotesque boundary violations.

Girard's argument against pre-Christian and Indigenous religions stems from his theory of what he calls "mimetic desire." He posits the innate capacity to imitate the needs and desires of others, or "acquisitive mimesis," as the clue to understanding human nature. While mimesis is a natural feature of the subject, it inevitably leads to tragic results by blurring distinctions and merging identities whenever the subject becomes successful in obtaining its object of desire. Now the mediator who had modeled a craving and attachment to the object becomes the rival who guards the subject from obtaining the object. Both parties see themselves in the other, imitating each other in a merging of their separate identities and a loss of the distinctions between self and other, model and disciple, that had once defined their relationship.

Historically, the gut-level response to the debilitating threat of unregulated desire is to turn a blind eye to the real cause of the problem—the raw compulsion to acquire the desired object—and impute to an unprotected "other" the cause of the community's dissolution into an undifferentiated and disordered state. The other now becomes the victim, the scapegoat, of the group's disintegration insofar as the other functions to divert collective violence to itself and away from the mimetic crisis. To legitimize the process of victimage, the group notices telltale signs of the other's destructive power, for example, its animal-like resemblances or some physical mark or deformity, as confirmation of its guilty status. The solution to mimetic crises, therefore, is the prophylactic of scapegoating violence. In order for the community to save itself from the inevitable corrosion of mimetic disorder, it must periodically plunge itself into a paroxysm of violence toward an abnormal and guilty scapegoat. Girard's final move is to say that while so-called primitive religion has its origins in sacrificial violence, the religion of Jesus in the Gospels repudiates such violence by uncovering the scapegoat mechanism at the base of culture and by promoting an ethic of love that allows all persons to expose the lie that scapegoating is inevitable and necessary.[31]

My approach to understanding this schema centers on Girard's analysis of the abnormal victims, often portrayed as deified mortals or sacred animal-like beings, who bear the marks of their victimage. He focuses on outcasts and rebels who are alternately the catalysts or the result of violating sacrosanct social norms. For Girard, healthy societies preserve traditional class, gender, and ethnic divisions. In contrast, unhealthy societies are contaminated by "slayers of difference" who undermine social distinctions in favor of an undifferentiated, Hobbesian war of all against all. His many examples of such persons include the godlike and "transvestite" Pentheus in Euripides, ass-headed Bottom in Shakespeare, and the divine dog-woman among Canadian Indians.[32] These bizarre creatures, who erase the differences between gods, humans, and animals, either are symbols of mimetic chaos or are seen as the perpetrators of chaos themselves. Girard gives many examples of these heteromorphic outcasts in ancient materials, biblical books, Amerindian stories, and medieval persecution texts, to name some, but let me illustrate the tenor of his thinking by noting three of the mythological figures from classical sources he uses to advance his theory.

In antiquity, Girard keys on Euripides' *The Bacchae* to highlight the power of mimetic frenzy to destroy differences and wreak destruction. In this tragedy, the god Dionysus inspires his female followers to become orgiastically insane, leaving their families to hunt and copulate with wild animals and satyrs in the forest outside the city walls. In turn, one of Dionysus' adoring followers, Agave, mistakes her son for a lion, killing him on the spot, and then presents his head to her beloved god as a prize of her sick blood sport. Girard also gives the example of Leda and the swan. Here Zeus is represented by a swan who seduces or rapes Leda in the woods, perhaps on the same night she sleeps with her human husband, and in some versions, Leda then lays two eggs from which some of her partly divine and partly mortal children are hatched. Similarly, Girard notes how Pasiphae, herself a crossbreed of Helios, the sun god, and a human mother, becomes besotted with the god Poseidon in the form of a bull, makes love with the divine bull in an open field, and then gives birth to the Minotaur, the monstrous flesh-eating hybrid creature with the head and tail of a bull and the body of a man.

All of these examples make the same point in Girard's theory: nature is the place where "monstrous couplings between men, gods and beasts are in close correspondence with the phenomenon of reciprocal violence and its method of working itself out."[33] Thus, mimetic helter-skelter takes place in a destructive and chaotic world in which the lines of division between the gods and humankind have collapsed; where gods, humans, and animals all fuse into one "muddy mass" of undifferentiated horror; where sexual norms give way to libidinal license; and where mortal and godlike women, Agave, Leda, and Pasiphae, travel outside the confines of civilization into untrammeled nature, where they violate sacred taboos, or are violated themselves, and thereby descend into bestiality and madness.[34] Consistently in Girard's writing, the natural world is a site of violence and anarchy—the place, for example, where Agave performs infanticide and Leda and Pasiphae make love with grotesque animal gods. Nature, in sum, is the locus of boundary-violating chaos in opposition to the normal order of structured and differentiated civilized societies. This binary opposition in Girard—the opposition between mimetic frenzy and social norms, between wild nature and cultural order, and between grotesque identity confusion and the ordered differences that separate divinity, humanity, and animality—is one of the primary structural dynamics that drive the meaning of his work.[35]

In my judgment, Girard's relegation of uncivilized wilderness, monstrous hybridity, and primitive religion to the natural world of blood and madness and his elevation of proper society, social order, and Christianity to the civilized world of safety and equilibrium falsely separates what biblical religion carefully blends and mixes together. Far from being an undifferentiated mass of confusion and violence, the natural world, and its many divine, human, and animal denizens, is the primary locale of the biblical God's revelation of peace and fecundity. *Above all else, nature is God's preferred habitat in the Bible.* In Genesis, God partners with the sacred ground to make human beings out of the fertile soil (2:1–9). In Job, God answers Job's theodicean cry with a plea for Job to look to nature for answers—and especially to the Behemoth, perhaps the great hippo, the first of all of God's works (40:15–24). In the Gospels, Jesus is born in a stable (Luke 2:1–20), uses agriculture as the basis of his many parables (Matthew 13), and grounds his most sacred teachings about baptism and Eucharist in the primal elements and foodstuffs of water, wine, and bread (John 3:1–10, 6:41–59). And also in the Gospels, as we have seen, God incarnates Godself in the animal body of a dove, conjoining the divine and animal orders into a kind of *theointerspeciesism*—the very transgressing of species lines against which Girard seeks to isolate nonsacrificial biblical religion (Matthew 3:13–17; Mark 1:9–11; Luke 3:21–22; John 1:31–34).

In the Bible, it is earthen wild places—it is the natural world in all of its glory and wonder—that is the interspecies body of God's revelatory activity. It is in nature where Moses encounters a burning bush—God as a sacred plant—and where God speaks of God's perfect identity and instructs Moses on his divine mission (Exodus 3:1–15). It is in nature where two lovers in the Song of Songs sing a rapturous poem of erotic delight: "Your rounded thighs are precious jewels, your breasts are palm clusters that I want to climb, and your kisses are the best wine that glides over my lips and teeth" (Song of Solomon 7:1, 8–9). And it is in wild nature where the Markan Jesus, symbolized by the lion, is ministered to by wild beasts and angels at the threshold of his public ministry (1:12–13); where the Johannine Jesus, as we have seen, makes mud pies as poultices to heal the man born blind from birth (chapter 9); where the Lukan Jesus goes to pray great drops of blood in the Garden of Gethsemane on the night of his arrest (22:39–46); and where Jesus' body, in Matthew's account, is laid to rest in a rock-cut tomb after a life spent in compassion and struggle

(27:57–61). In the Bible, contrary to Girard, the natural world, and its wealth of flora and fauna, its bounty of gifts and sustenance, its ever-flowing offerings of grace and magic and beauty—it is the natural world itself, far from being a dangerous melee of demigods and monsters, that is the privileged site of God's power and habitation for all of God's many and diverse children.

## *Green Mimesis*

In Chapter 1, I said that while the first naivete of primordial animism is lost for many of us, now the second innocence of biblical animism is available as a viable worldview sufficient for rekindling our spiritual bonds with our plant and animal relations. And in this regard, Girard's mimetic theory is crucially important for understanding the attitudinal gesture necessary for reforging primal connections with our earthbound cousins.

Girard's critique of ancient religion, and corresponding vision of nature as a muddle of violent identity confusion, is wrongheaded, I have argued. But his notion of mimesis is a productive idea for cultivating the right emotional framework for healing our deformed relationship with our planetary habitat. Girard maintains that there are actually two modes of mimetic expression that define the human condition: *acquisitive mimesis*, which leads to rivalrous chaos, as we have seen, and also *nonacquisitive mimesis*, which imitates the healthy desires of others and does not descend into a whirlpool of violence and retribution. He writes, "On one side are the prisoners of violent imitation, which always leads to a dead end, and on the other are the adherents of non-violent imitation, who will meet with no obstacle." At another point, he flatly declares, "Mimetic desire is intrinsically good."[36] His point is that while mimesis can easily degenerate into rivalry and aggression, mimesis can also lead to positive identity formation as the subject learns to appropriate the other's desires while still maintaining thoughtful boundaries between self and other. The goodness of mimesis, as Girard puts it, inheres in the subject's capacity to grow and develop through nonrivalrous imitation of the other without the need to acquire or dominate the other in the process.

Nurturing healthy mimetic relations with our animal and plant relations, in my judgment, begins with finding Spirit refracted across the spectrum of commonplace existence. In the history of Christianity, however, this refraction is often missed within the landed reality of God's pedestrian presence in everyday life. What is forgotten is that it is precisely *within*—not *outside*—the beating heart of the natural world that God is set forth. What is not noticed is that God is right here, present in ordinary existence, veiled but still apparent in the sure flight of a barn swallow in the spring twilight, the soft touch of a baby's grip in a grandparent's rough hand, the heady aroma of fresh basil in an early fall garden, the taste of salt and sweat on lovers' skin in the sticky heat of late summer. There is no place else to look, there is no place else to go, there is nothing else to see or hear or taste or feel. Moreover, there is no temple or holy man or sacred book that is required to traverse the boundary of heaven and Earth, to make God come into this world, to realize God's presence through this or that religious ceremony or the agency of a special mediator. Why? Because, according to scripture, God is always-already here—present, tactile, enfleshed, and enlivened by all that is around us. Jesus makes this point especially clear in his teaching about the kingdom of God even now being present in the everyday. When he is pressed by religious leaders in Luke's Gospel concerning the whereabouts of God's kingdom, we read, "Now having been questioned by the Pharisees as to when the kingdom of God was coming, [Jesus] answered them and said, 'The kingdom of God is not coming with signs to be observed; nor will they say, "Look, here it is!" or, "There it is!" For behold, the kingdom of God is in your midst'" (Luke 17:20–21). The kingdom of God is here and available for all to enjoy: touch, taste, hear, and see God in the wondrous droplets of common grace always and everywhere present in ordinary time and the familiar of routine but charmed existence.

In this chapter, I have sought to put forth Earth as an all-encompassing gift with the potential to heal and restore all of its inhabitants—an enchanted place where seen and unseen forces can provide daily sustenance, pleasure, and renewal. Christians have often longed for the arrival of the Spirit in nature—a longing crystallized in the ancient Christian cry, "Come, Holy Spirit, and renew the face of the Earth!"[37]—in order to realize God's power to quicken and sustain everything that God has made. This cry is analogous to the *epiclesis*, the ancient invocation of the Eastern Orthodox churches in which the priest "calls down" the Spirit to be present in the community. This longing for the Holy Spirit—the healing and vivifying force who brings to life all things—defines all Christian experience of God's presence in the world. In my life, this hope and desire is for the Spirit to make real the naturally supernal wonders that weave together the mystery and joy of creation—including, in my case, the beauty of the wood thrush, the vitality of the Endless Mountains, the ancient dirt and spit at the Siloam pool—and, as I will note in the next section, the jubilant, raucous arrival of a particular family of birdlife in the marvel that is the Crum Woods.

Sometimes in the Bible, the Spirit is figured as a veiled force, present within all things but unheralded, like the physical element of air that we need for survival but often take for granted. This is the case in John 20:21–22, which says, "Jesus said to [the disciples] again, 'Peace be with you. As the Father has sent me, even so I send you.' And when he said this, he breathed on them, and said to them, 'Receive the Holy Spirit.'" This passage revolves around a conceptual pun. In the Greek New Testament, the noun for Sprit, breath, or wind (*pneuma*) and the verb for breathing-upon (*emphusao*) both stand for the same reality: the reality of breath, the act of aspirating. To capture this wordplay, John 20:21–20 instead could be translated as "And Jesus *breathed* on them and said, 'Receive the Holy *Breath*'" or "And Jesus *aspirated* on them and said, 'Receive the Holy *Spirit*'" or "And Jesus *blew* on them and said, 'Receive the *Wind* of God.'"

Here the Spirit, while not seen as such, is the life-giving reality of the power to breathe—to be a human person is to take God as Spirit into oneself by mouth, as it were, through the simple act of inhaling and exhaling the gift of fresh air. At the start of our lives, we breathe God in; at the end of our lives, we breathe God out. To be a living being on the planet is to inhale and exhale God's Spirit—the wind of God, the breath of God—in harmony with the great "oxygen cycle" that makes Earth capable of sustaining life.[38] The Holy Spirit, appositionally speaking, is the airy atmosphere itself, the breath of God that aspirates, or in-spires, all life—vivifying all things, empowering all things, making all things dance into existence through the living dynamism of God's earthly, breathy presence.

But just as often, the Spirit is vibrantly depicted in concrete and tangible form, as we saw in the case of the bird-God in Luke's account of Jesus' baptism or as tongues of fire on the disciples' heads in Acts' account of the day of Pentecost. Where, then, are the physical intimations of God's Spirit in the world today? Like the disciples on the road to Emmaus in Luke 24, who walk and talk with Jesus but do not recognize him for who he is, is it not also that God is corporeally everywhere in our midst as well but that we do not recognize the many faces of the divine alive and in force in our everyday comings and goings?

## *The Pileated Woodpecker*

This spring in the Crum Woods—on the southern edge of the Delaware River Basin, two hundred miles from the Endless Mountains, Carol

French's farm, and the gas fields of the Marcellus Shale—I encountered a wide variety of birds in my morning walks with my friend Dave Eberly. Here is a list of the different types and numbers of species we saw and heard on one typical spring morning:

one mallard
twenty double-crested cormorants
one broad-winged hawk
eight mourning doves
seven red-bellied woodpeckers
four downy woodpeckers
one hairy woodpecker
five northern flickers
one pileated woodpecker
one Merlin falcon
six blue jays
two American crows
two fish crows
eight tree swallows
six Carolina chickadees
ten tufted titmouses
one white-breasted nuthatch
one Carolina wren
sixty American robins
one northern mockingbird
four song sparrows
seven white-throated sparrows
one dark-eyed junco
five northern cardinals
eight common grackles
two brown-headed cowbirds
four house finches
four house sparrows

Amid this display of avian biodiversity, I became especially intrigued with the growth and the life cycle of a pileated woodpecker family. The pileated woodpecker is a strikingly beautiful winged creature. It is a large bird, about the size of a crow, mostly jet-black with slashing white stripes and a bright red crest on top of its head. I first saw one of the mature woodpeckers hammering out a nest hole in a dead tree on the banks of the Crum Creek. It cut wood chips out of the tree and then flicked them away from the newly excavated cavity with relish abandon, many of them splashing into the creek below. I watched this bird and its companion often feeding below the nest—pileated woodpeckers like ants and termites—by banging repeatedly into nearby trees. This is a loud, brash, in-your-face bird that lights up the forest with its soaring flight, screeching call, and head-bashing impact. After a couple of weeks of nest building by the adults, wonderfully, two baby pileated woodpeckers appeared. They screeched and clattered around the nest just like the parents. Another two weeks passed, and the babies fledged—effortlessly, it seemed, as they took flight through the forest echoing their parents' wild energy and noisy calls.

I was impressed with the care and concern the adult woodpeckers showed their young. Building a nest cavity of perfect proportions, soaring through the forest and chopping into trees for grubs for their babies, and huddling together in the nest day in and day out—I thought to myself, "This is how I want to care for my own young adult children, Katie and Chris." Creating a home, providing resources, offering encouragement—in essence, teaching my children to fly. When Katie and Chris were younger, we often went into the Crum Woods for passive recreation—sitting on a rocky outcrop, looking for birds, swimming in the creek—and I felt then that immersing oneself and one's family in wild places is necessary for human flourishing.[39] On this point, I find Girard to be very helpful. In Girardian terms, modeling my life with my family after the woodpeckers and their kind is mimesis at its best. My hope is to live my life as an homage to the pileated woodpecker family—to enjoy my life with my children, as the parent birds did with theirs, in my modern-day analogues to the woodpeckers' rituals of hunting, gathering, and nest building. My ideal is to shadow the nesting woodpeckers as an exercise in "green" mimesis, an *imitatio naturae*, in which I learn to emulate the woodpeckers' singleness of purpose by cultivating my own loving and intimate household as well—just as my aerial friends have shown me.[40]

This reflection returns me to my question throughout this book: is God in the face of the pileated woodpeckers I find overhead in the Crum Creek? In response, a historical note: In the past, the pileated woodpecker's closest cousin was the ivorybill woodpecker. The two woodpeckers are almost indistinguishable, with the ivorybill being a bit larger than the pileated with an ivory-colored bill (so its name) as opposed to the pileated's gray and whitish bill. But in spite of their almost identical appearance, the ivorybill is now considered a "ghost bird," a bird that is most likely extinct, because its original pre-twentieth-century habitat, the swampy primeval forests of the American Southeast, was largely cut down for timber production.[41] However, its counterpart, the pileated woodpecker, is thriving in the ivorybill's home range and beyond, in large part because of modern sustainable forestry practices, so much so that the pileated woodpecker now flourishes in a large woodland arc extending from the eastern United States through Canada and into the Pacific Northwest, including Northern California.

In the Crum Woods, I was spellbound when the pileated woodpecker would leave the nest and soar through the trees, darkening the ground at my feet with its almost-thirty-inch wingspan. Its head-banging sound would ricochet through the forest; its flight through the leaves and sky overhead would root me in place, unbelieving that such a creature is still with us. The ornithologist Scott Weidensaul writes about the similar reactions local people have experienced over time when the ivorybill, in particular, and the pileated woodpecker as well have taken flight above their heads. Historically, this spectacle prompted observers to call these sensational creatures the "By God, look at that bird" woodpeckers—or, simply, in moments of gasping, exclamatory wonder, "the Lord God bird." Weidensaul writes, "The names [country folk] used for the ivorybill reflected the dazzle of seeing one of these huge birds rowing through the light-splashed swamps on powerful wings. King of the Woodpeckers, they called it. Log-cock. King Woodchuck. Giant woodpecker. Log God. Like the smaller but similar pileated woodpecker, it was sometimes called the Lord God bird, or the By-God, because that's what a breathless greenhorn said when he first saw one: By God, look at that bird."[42]

To be sure, such outbursts can be read simply as impulsive reactions to the sudden appearance of a large, multicolored bird ripping apart old logs with its massive bill or flying through the forest, blackening out the sun's light overhead. But can such outbursts at the sight of the ivorybill or pileated woodpecker also be read as a hint of something else? Is it possible such outbursts are a clue in our core psyches that we are aware of something awe inspiring—even divine—at play in the enchanted forests that surround us? Could such outbursts signify a rumor within our deepest selves that we are being visited by powers greater than ourselves—powers that offer us meaning and courage in a world increasingly devoid of hope in the future well-being of the planet?

For me today, crying out "Lord God!" at the sight of a pileated woodpecker, as others did at the sight of an ivorybill or pileated in days gone by, is a shout-out to all of creation that here in this place and at this time I am encountering a visage of God that is not unlike, I suspect, the reaction of

Jesus' onlookers to the aerial Spirit above them at the time of Jesus' baptism in the Jordan Valley. This, then, is my heartfelt belief and my hope: catching a site of the Lord God bird in the Crum Woods is to catch a site of God flying and rustling in my midst. In my "Lord God!" exclamation, my belief and my hope is that God is still active and alive in the woodlands and fields and rivers within all of this beauteous creation. My belief and my hope is that the winged God at creation, and again at Jesus' baptism, is once more taking flight as the Lord God bird in the Crum Woods, the avian God of the forest, the pileated woodpecker, who again calls to me and to all of us, from the byways and the hedges of our daily lives, to find and pursue grace and justice in our place and in our time as well.

CHAPTER 3

# Worshipping the Green God

### *Crum Creek Visitation*

I recently encountered a great blue heron—similar in splendor to the pileated woodpecker—while teaching my Swarthmore College class Religion and Ecology. I was conducting a three-hour class meeting in the Crum Woods, the scenic watershed adjacent to the Swarthmore campus and near my home. The class began with a silent procession into the woods, where I asked each student to experience being "summoned" by a particular life-form in the forest—blue jay, gray squirrel, red oak, water strider, skunk cabbage, and so on—and then to reimagine ourselves as "becoming" that life-form. After the walk through the woods, we gathered in an open meadow, under the shade of a grove of sycamore trees, so that each student and I could "speak" in the first person from the perspective of the individual life-form we had assumed.

I explained that this was a voluntary exercise and that no one should feel compelled to speak if he or she did not want to. I said, "If you imagine yourself, for example, as a brook trout or mourning dove or dragonfly living in and around the Crum Creek, with the creek threatened by suburban storm-water runoff, invasive species, and other problems, what would you like to say to this circle of human beings? This group activity is a variation on a deep ecology, neopagan ritual called "A Council of All Beings," in which participants enact a mystical oneness with the flora and fauna in an area by speaking out in the first person on behalf of the being or place with which they have chosen to identify.[1] A Council of All Beings ritual enables members of the group to speak "as" and "for" other natural beings, inventively feeling what it might be like to be bacterium, bottle-nosed dolphin, alligator, old-growth forest, or gray wolf. A Council is an exercise in imaginative ontology. Participants creatively metamorphose into this or that animal or plant or natural place and then share a message to the other human persons in the circle. The purpose of a Council, then, is to foster compassion for other life-forms by ritually bridging the differences that separate human beings from the natural world.

Trained in religious studies, I was mentored to avoid leading contemplative practices in the classroom lest students confuse the academic study of religion with particular sectarian rituals. It is one thing to study, say, Christian monasticism as an intellectual exercise, so the argument runs, but quite another to practice the daily office as a spiritual exercise. But while I think it is wrong to try to inculcate particular religious beliefs or practices in the classroom environment, I have found my students increasingly hungry not only for religious studies as a mode of *critical* analysis but also for the actual *lived* experiences that underlie and shape such studies. Thus, I use contemplative rituals with students as integral to their academic inquiries. These exercises affectively ground the analytically discursive work of writing papers and taking exams in my classes. Now both modes of learning—heartfelt mindfulness and academic analysis—provide the sonic baseline that shapes the rhythms of my pedagogy. In this way, I conceive of my teaching as a type of *soulcraft*. My goal is not simply to communicate intellectual content, as important as that is, but, moreover, to facilitate conversation and contestation about transformative wisdom traditions grounded in affective practices and ritual formation.

So on this particular day, as I and my students were reworking our identities and speaking as new life-forms, a great blue heron broke the plane of blue sky above our heads and glided effortlessly toward the creek. We were spellbound. Flying with its neck bent back in a gentle horizontal *S* curve, its blue-black wings fully extended, and its long gray legs ramrod straight and trailing behind, the heron darkened the sky above our heads and landed in the shallow water of the creek. It was as if a pterodactyl from 150 million years ago had entered our gathering. The slow beat of its great wings—whomp, whomp, whomp—followed a stately rhythm from another era. Spontaneously, we jumped up from the meadow and walked heron-like—silent, hands held at our sides, strutting in the tall grass—toward this imperial creature, who began to stalk and then strike its prey with its long yellow bill.

*This is a visitation.* My class and I now felt, it seemed to me, that our modified neopagan shape-shifting practice had prepared us for a connection—a spiritual connection—to a regal bird whose presence electrified our gathering with agile dignity and rhythmic beauty. We had been discussing, and trying to ritually enact, our identities as fellow and sister members of this forest preserve in communion with the other life-forms found there. But to be graced with the overhead flight and practiced movements of the great blue heron transformed, at least for me, what we had been learning and practicing into a living relationship with a kind of forest deity, if I could be so bold. Together on the banks of the creek, we stood and watched the heron balance itself on one leg, silently step toward its quarry, jerk its bill toward an unsuspecting fish or frog, strike and swallow, and then rise in flight again off the creek, its great V-shaped wings flapping in slow-moving unison. I am not sure of my students' experience at this point, but to me, I felt wonder and gratitude in the presence of this awe-inspiring creature, remembering Moses' experience of God in the burning bush: "Come no closer! Remove the sandals from your feet, for the place on which you are standing is holy ground!" (Exodus 3:5).

In our outdoor class, we had been discussing how our lives were embedded in a sacred hoop greater than ourselves, that as human citizens of a wider biotic community, we are surrounded by a cloud of witnesses who are calling us to our responsibilities for preserving the woods, preserving ourselves, preserving the world. Now I sensed my class, and I knew what was at stake in this preservation. What was at stake was the preservation of this beautiful forest refuge—the Crum Woods, symbolized by the blue heron on the wing—as a biologically rich home for it many denizens, including its human visitors who enjoy the forest from time to time for its many sensual and meditative delights. Whatever my students' experience of the great blue heron was that day, for me, this encounter underscored my conviction that the Crum Woods is more than a biodiverse habitat; it is also, in my Christian vocabulary, a green sanctuary, a blessed community, a sacred grove, indeed, a holy place.

But, as we have seen, calling the Crum Woods, or any landscape, a *sacred place* is dangerously transgressive for the guardians of Christian orthodoxy who circumscribe the boundaries of Christian belief in the divine life to God alone. Such critics consider it idolatrous to open up the language of holiness or divinity to include the natural order of things. For example, the Protestant ecotheologian John B. Cobb Jr., who emphasizes God's presence in creation, appears wary of extending the horizon of the sacred to encompass the more-than-human world. He writes that Christian thinkers can "celebrate the growing sense of the sacredness of all creatures," but he then seems to back away from this claim, saying,

> Nevertheless, that language is, from a historic Protestant perspective, dangerously misleading. Speaking rigorously, the line between the sacred and the profane is better drawn between God and creatures. To place any creatures on the sacred side of the line is to be in danger of idolatry. . . . God is present in the world—in every creature. But no creature is divine. Every creature has intrinsic value, but to call it sacred is in danger of attributing to it absolute value. That is wrong.[2]

Another ecotheologian, Richard Bauckham, makes a similar point. While he extends the horizon of sacrality to the natural order of things, pace Cobb, he nevertheless contends that mainstream biblical teaching compels him to stop short of ascribing divinity to anything other than God. "The Bible has de-divinized nature," he writes, "but it has not desacralized nature."[3] Bauckham steps beyond Cobb's polemic against sacred nature, but then he balks at the thought of attributing the attributes of God in Godself to the creaturely world as an expression of pagan animism foreign to the tenets of what he calls the "Judeo-Christian tradition." He continues, "The Judeo-Christian tradition certainly de-divinized nature. Nature is not reverenced as divine in a pantheistic or animistic sense. But deeply rooted in the Judeo-Christian tradition is the sense that all creatures exist for the glory of God and reflect the glory of God."[4]

Both Cobb and Bauckham rightly celebrate God's presence in nature insofar as all things are reflections of God's goodness and glory. They contend that nature should not be denigrated in utilitarian terms as an object of exploitation but valued instead as an expression of divine beauty and providence. Nevertheless, both authors' anxieties about divinized nature and pagan animism move them away from the model of Christian animism proposed here. Perhaps their caveat against ascribing sacred value (Cobb) or the quality of deity (Bauckham) to the creaturely world stems from their reading of Paul's invective against pagans who "exchanged the glory of the immortal God for images resembling mortal man or birds or animals or reptiles. . . . They exchanged the truth of God for a lie and worshipped and served the creature rather than the creator" (Romans

1:23, 25). But while this passage has traditionally been used to empty the world of signs of divine presence, it is important to note the historical context of Paul's comments. Paul says it is wrong to worship God in the image of mortal humans and other beings, but what exactly does this mean in reference to Christianity's core belief that God became flesh and should be worshipped as such accordingly?

In spite of Paul's polemic against exchanging reverence for God for idol worship of a man, I do not think he is saying that Christians cannot glorify God in human and material form, because that is exactly what Christianity is—a religion that worships the manifestation of God in the flesh of a mortal man, Jesus of Nazareth, and, as I have maintained, also reveres the embodiment of God in the flesh of a mortal animal, the Holy Spirit. As well, this worship takes on many other earthly and creaturely expressions, in Paul's time and in our own time—from devotion to icons and veneration of Mary to blessing the host and giving precious metals and candles to saints as votive offerings. In the light of Christianity's foundational Trinitarian-incarnational identity, it is most plausible, then, that Paul is not inveighing against Christians who worship God in the form of mortal creatures or the material world. Rather, he is arguing against people outside the circle of biblical faith who divorce their religious devotion to images or other creaturely manifestations of God from the understanding of the enfleshment of God in Christian tradition. In Christian terms, and I think Paul would agree, it is not wrong to worship God in corporeal form, but it is wrong to do so and not celebrate such worship as giving praise to the God of the biblical witness who makes and sustains all things.

My point is that a thoroughgoing incarnational model of Christianity sees no division between the God of the biblical texts, on the one hand, and the sacred and divinized character of creation, on the other. Indeed, the two affirmations mutually support each other. In this regard, dialectically speaking, God and nature are one. Here I have sought to show that Christianity is a faith that celebrates the embodiment of God in many forms—and not only in human form in the person of Jesus but also in animal form in the person of the Spirit. Christianity, as I have suggested, is a religion of double incarnation: in a twofold movement, God becomes flesh in both humankind and otherkind. Just as God became human in Jesus, thereby signaling that human beings are the enfleshment of God's presence, so also by becoming avian in the Spirit, God signals that other-than-human beings are also the realization of God's presence.

To suggest that Christian faith, at its core, centers on belief in God as a fully incarnated reality not only in the humanity of Jesus Christ but also in the animality of the Holy Spirit means that the embodied Spirit of God sets forth the whole wide-ranging world of nonhuman nature—the birds of the air, to be sure, but also the fish of the sea, the beasts of the field, the trees of the forest, indeed, all of creation—as saturated with divinity. And my practical point is that the full realization of this truth should lead all of us to comport ourselves toward the natural world in a loving and protective manner because this world is the unbounded fullness of God within the life of each and every created thing.

## *Christian History*

In this chapter's opening vignette, I profiled my encounter with the great blue heron's feeding dance, along the banks of the Crum Creek, as a reminder of the presence of God in the midst of everyday, natural life. Watching the heron subsist in balance with its neighbors in the Crum watershed turns my gaze to the sacred in my midst and puts my feet on holy ground. But throughout this work, I have been quick to point out that this affirmation of carnal divinity has historically divided some Christian thinkers from others. Paul Santmire has shown that the emphasis on God's embodied presence in all things is in constant historical tension with a counterhistory that accentuates God's distance from and, at times, hostility toward the ecological order of things. Santmire writes that traditions that stress God's here-and-now incarnational power within creation revolve around what he calls the "metaphor of fecundity," while traditions that drive a wedge between God and the world, by highlighting the supreme value of the spiritual world above and beyond this material world, focus on what he calls the "metaphor of ascent." These two metaphors vie with each other for importance throughout Christian history. As Santmire puts it, "If the metaphor of ascent is dominant, then the metaphor of fecundity—insofar as it is influential—will be systematically subordinated to the metaphor of ascent. Categorically, the overflowing goodness of God will be viewed as the first stage in a universal divine economy whose final goal is the ascent of the spiritual creatures alone to union with God."[5]

I am persuaded by Santmire that these two dynamics—ascent and fecundity—are the root metaphors that enliven classical Christian understandings of the natural world. The metaphor of ascent, on the one hand, regards creation as a way station in the soul's upward journey away from this world and into union with God in the world beyond. The ethical import of this metaphor is that Earth is irrelevant, or even inimical, to the work of human salvation and can thereby be disregarded or exploited by its human overseers. The metaphor of fecundity, on the other, champions creation as the good land that God fills with God's presence and gives to all beings for their sustenance and well-being. Ethically, this metaphor brings to life a stewardship model of human beings as caretakers of God's earthly bounty.

Glossing Santmire's categories, I will analyze in what follows the *animistic* and *otherworldly* dispositions in a variety of Christian and spiritual thinkers and traditions. Like Santmire's study of the metaphors of fecundity and ascent, I will oscillate between examining world-affirming and world-denying modes of understanding with particular reference to the degree to which the thinker or tradition in question articulates an animist sensibility. My aim is to survey Christian history with an eye toward the importance of *divine enfleshment* as the animating principle by which the New Testament and many central Christian thinkers have understood the meaning of biblical faith. I do this in order to provide the classical warrants for Christian animism even in cases where some of the traditions and thinkers I examine do not obviously fall, at least at first blush, into this theological paradigm. My hope is to gather together a cloud of witnesses to the belief and experience of sacred, divinized nature within the Christian scriptures and subsequent thinkers in the Western theological tradition. To this end, I will study here Jesus of Nazareth, Augustine of Hippo, and Hildegard of Bingen, and in Chapter 4, I will take up John Muir. My goal is to bundle together spiritual traditions and thinkers that envision the green mantle of Earth community as a site of awe and wonder—indeed, as the place of God's full presence and daily habitation. I begin in the beginning by interpreting the Gospel narratives of Jesus' interactions with the natural world as the founding story line in my Christian animist proposal.

## *Jesus and Sacred Land*

Making sense of Jesus' life through the optics of consecrated nature, one sees that he derived his strength from his deep-seated identity with particular landscapes charged with divine energy. As an itinerant preacher and healer, Jesus was an intimate of wildlife and natural settings where the power of divinity was keenly felt and where that power could be channeled for the healing of others. As discussed in Chapter 1, he began his public life in the Jordan River, where the bird-God emerged from a hole in the sky and protectively landed on him during his baptism by John. Leaving the river, the same bird-God drove him further into the wilderness, where he lived with wild animals and angels who cared for him. Emerging from his fasting and praying in the countryside, Jesus began to teach that God is the creator and lover of all things, including the birds of the air, who are nourished by their heavenly father (Matthew 3:13–4:2, 6:26).

Every generation invents Jesus afresh according to its own cultural presuppositions. Indeed, as Albert Schweitzer said, every generation that seeks to discover the historically authentic Jesus of Nazareth only finds itself looking down the well of its own imaginings—and, thereby, mistaking its own face looking back at itself for the authentic face of the Jesus of history.[6] No generation finds the true Jesus, only its own self-reflection in the deep well of history. Be this as it may, what most generations seem to have missed is that whatever else Jesus was, he was, first and foremost, a traveling teacher and healer who himself, it seems, and as he instructed others to do, lived off the land with no home, no money, no prospects, and apparently, not even a change of clothes. Gathering together his disciples, Jesus "said to them, 'Take nothing for your journey, no staff, nor bag, nor bread, nor money—not even an extra tunic'" (Luke 9:3). Jesus was rich in human partners, animal friends, sacred plants, and hallowed landscapes that fed his spirit, provided him with rest and security, and empowered his demanding ministry to others. Jesus, then, did not define himself according to his material possessions. Rather, it was his radical identity with the land—and its many strange and wonderful denizens—that is the defining trait of his personality and ministry.

Jesus has been seen as a traveling philosopher, a political revolutionary, and a religious reformer. All of these descriptions are appropriate within their particular spheres of interpretation. What is missing, however, is that Jesus, in addition to being a teacher, was essentially a healer—more particularly, a *shaman*—who interacted with natural elements as the means by which he was able to restore health and well-being to his followers and supplicants. Utilizing social-scientific tools to account more fully for the shamanic features of Jesus' public life, the anthropologist Pieter Craffert situates Jesus' roles as healer, exorcist, and spirit-medium within the vernacular religious culture of his time. He writes, "Notions of controlling the elements, experiencing spirit-possession, controlling and communicating with spirits, experiencing miraculous healings, recounting special births, and the like are stories that make sense in many traditional cultural systems and particularly in a shamanic worldview. For Jesus' compatriots, Jesus could control the elements, walk on the sea, provide food and drink, converse with the ancestors, return after his death, and so forth."[7] I think Craffert is correct in his interpretation of Jesus' shamanistic interventions into the natural world and use of material elements to perform healing rites. Instead of "controlling the elements," however, I prefer to say Jesus mediated or marshaled the elements in order to accomplish his medicinal ends. His practical goal was therapeutic and curative, not mastery and control. In any event, it is important to note that Jesus preferred particular natural elements—namely, touch, saliva, and oil—as the holy means by which he was able to channel God's healing and renewing power to others.

Steeped in the restorative powers of the living earth, Jesus conducted divine energy to heal and enliven his followers and others seeking wellness. In particular, healing touch and saliva therapy were central to his mission. When a leper knelt before him, Jesus stretched out his hand to touch him and free him from his disease (Matthew 8:1–4). When a blind man was brought to him, Jesus spat in the man's eyes, laid hands on him, and restored the man's vision (Mark 8:22–26). For another blind man, as we saw in Chapter 2, he made a wet earthen plaster with his spit and then patted the muddy compress onto the man's eyes, also restoring his sight (John 9:1–12). Animals were part of his therapeutic regimen as well. In the case of the demon-possessed man who cut himself with stones, Jesus this time did not rely on therapeutic touch or spittle to heal the man but instead turned to a nearby herd of swine as the repository of the man's exorcised demons—sending the demons into the pigs, who then raced off a cliff and killed themselves below (Mark 5:1–20). Sometimes relying on his own bodily modalities and at other times utilizing animal bodies, Jesus sought to heal the disabled and the sick.

Medicinal oils—oils fragrant and infused with healing properties—also played a central role in Jesus' caring attempts to restore his disciples' and patients' potential for wholeness and balance when these capacities had been degraded by illness, injury, and loss. Jesus was given the curative spices of frankincense and myrrh at his birth (Matthew 2:7–12); he was anointed and refreshed by Mary of Bethany at a dinner party with a flask of pure nard, a costly perfume imported from the Himalayas (John 12:1–8); and most significantly, he instructed his disciples to visit all of the villages in the Galilee, where he had grown up, to teach, exorcise, and heal, and, according to the Gospel of Mark, they did so, preaching that all people "should repent, and they cast out many demons, and anointed with oil many that were sick and healed them" (Mark 6:7–13). All therapeutic oils are plant extracts; it was these plant-based emollients or liquids that the disciples, upon Jesus' urging, used to heal the sick. While we do not know whether Jesus himself used essential oils, we know he commanded his disciples to do so, directly or indirectly. Significantly, then, it was not just bodily touch or saliva that Jesus considered to be sacred or the bodies of animals he used for exorcism, but he considered many plants to be sacred as well. It was then—as it is now—the natural oil from these plants that served as remedies for fatigue, fever, viral infection, headache, flu, skin problems, and much more.

Jesus' shamanism—his ability to channel divine power through the natural world—grew out of his founding encounters with Spirit-filled landscapes and creatures in their everyday habitats. These encounters indelibly shaped his subsequent travels and teachings. He taught in parables having roots in the sacred earth and in which animals and plants were experienced as the bodily form of divine reality. Now creation itself became the medium of his daily discourse. He compared the kingdom of God to a mustard seed that grows into a great shrub (Matthew 13:31–32), a collection of seeds scattered onto the ground with only some surviving (Matthew 13:1–9), and a net thrown into the sea that catches a wide variety of fish (Matthew 13:47–50). He used agrarian imagery to identify himself to his followers: *I am the good shepherd who goes after the lost sheep* (John 10:10–12), *my flesh is the living bread I give for the life of the world* (John 6:51), *I am the true vine and my followers will bear much fruit* (John 15:1–5). And rhetorically, Jesus shape-shifted into animal forms in order to highlight his special mission: *I am the mother hen who gathers her Israelite children under her wing* (Luke 13:34), *I am the serpent in the wilderness who gives eternal life* (John 3:13–15), and, as was said of Jesus by John the Baptist, *behold the lamb of God who takes away the sins of the world* (John 1:35–37).

Jesus' love of the animal and plant worlds propelled him to frequent isolated backcountry areas—what the Bible calls "deserted" or "lonely" places—as a source of strength for his daily engagements with others. He needed time alone by himself, and he would often travel to uninhabited regions for rest and renewal. Perhaps these times apart were inspired by his inaugural sojourn into the wilderness, initiated by the bird-God, when he had lived with wild animals and angels. "In those days Jesus came from Nazareth of Galilee and was baptized by John. And just as he was coming up out of the water, he saw the heavens torn apart and the Spirit descending like a dove on him. . . . And the Spirit immediately drove him out into the wilderness forty days, tempted by Satan; and he was with the wild beasts; and the angels waited on him" (Mark 1:9–10, 12–13). Moved by the Spirit-bird into the countryside, it is not surprising that Jesus seemed especially drawn to secluded mountainous areas. Sometimes he would go into the mountains and spend the whole night there praying to God (Luke 6:12–16). At one point, he spent the day teaching in the temple, and afterward he camped on the Mount of Olives, where he slept through the night (Luke 21:37–38). It seems that Jesus had a special fondness for this particular mountain site—a holy place where the reality of God was especially potent. It was at the Mount of Olives that he came and went as was his custom, indeed, so much so that when the time came for his arrest, his enemies knew exactly where to find him—in a garden called Gethsemane nestled within the Mount of Olives (Luke 22:39–46).

My point is that the baseline of Jesus' life was his daily dwelling in what Celtic Christians call "thin places"—that is, natural or sometimes built settings where the dividing line between "heaven" and "earth" appears to be thinned out, if not erased altogether.[8] Today, to conjure the itinerary of the thin places Jesus frequented long ago—Sabbath grain fields, Judean desert, Jordan River, Sea of Galilee, Jerusalem Temple, Nazareth synagogue, Garden of Gethsemane, Jericho road, and Mount of Olives—is to hear the ancient musical cadence of his determined wanderings. Taking comfort and finding God in these thin places was the living ground tone of Jesus' labors, the places set apart by virtue of their centering influence in his life.

Jesus' mission emerged, therefore, from his deep communion with landed places integral to his daily peregrinations. His identity with spiritually saturated thin places entailed as well his sense of belonging with trees and flowers, wind-blown seas, and the starry atmosphere above. To take one arboreal example, Jesus knew all about the life cycle of the fig tree. Gathering his first disciples in Galilee, he recognized Nathaniel under a fig tree (John 1:48); he likened growth in faith to slow-growing fig trees, which take years of painstaking cultivation before they bear fruit (Mark 11:11–14); and he recognized the tender leaf buds of spring fig trees as portents of summer's arrival, which he compared to the coming of the Son of Man (Matthew 24:32). Jesus loved flowers. He invoked the everyday splendor of the "lilies of the field"—an echo, perhaps, of the bride's exultation in the Song of Solomon, "I am a lily of the valleys" (2:1)—when he said that these wildflowers "neither toil nor spin, yet Solomon in all of his glory was not clothed as beautifully as these lilies" (Matthew 6:28–29). Jesus had a deep familiarity with the Sea of Galilee, its tides and rhythms, and the sea life and ideal casting angles for fishing therein. In one scene, he nonchalantly fell asleep amid his agitated disciples in a boat being swamped by high winds and violent chop (Matthew 8:23–27); in another, he told his disciples how best to cast their nets to catch sea fish, and then he made breakfast for the fishing party (John 21:1–15). And he marveled at the heavens when he would sky watch for signs that would auger the meaning of a darkening sun and moon and falling stars at certain times of the year (Luke 21:25–28).

Birds in particular were everywhere in Jesus' life. Most likely because of the support offered to him by the bird-God at his baptism, Jesus looked to birds as spirit-guides, or totems, in his outreach to others. When a religious leader offered to follow him anywhere, Jesus cryptically replied that foxes have holes, and birds have nests, but the Son of Man has nowhere to lay his head (Matthew 8:20). God provides habitat for birds and other creatures, but where are Jesus and his would-be followers to find shelter without a home of their own? Significantly, he discussed three different species of birds—the rooster, the raven, and the sparrow—as totem-beings who can lead wise seekers into a deeper relationship with God by following these birds' laudable character traits.

In the first instance, Jesus understood that the rooster, or the cock, appears to be governed by its own circadian rhythms and that without fail, it will blast its cock-a-doodle-doo song each and every morning. On the night of his arrest, he understood this pattern and said to his follower Peter that before the cock crowed the next morning, Peter would deny that he knew Jesus—and not once but three times. In fact, Peter did deny Jesus, and the cock, as predicted, crowed its early-morning call (Matthew 26:69–75). I think the message of the story is that nonhuman beings in their natural habitats are faithful to their core selves and deepest instincts, whereas, as human beings, we are not. Indeed, unlike the birds and animals that stay true to themselves, we, like Peter, say one thing and then do another. As humans, we can be feckless and self-deceptive, and so Jesus is asking us to look to the other animal and plant life-forms that surround us as models of integrity and consistency.

This theme is consistent with Jesus' similar comments about birds other than the rooster. His affection for the raven is a second example of his avian predilections. "Consider the ravens," he says in the Gospel of Luke; "they neither sow nor reap nor gather into barns, and yet God feeds them" (12:24). But as human animals, we no longer "consider" the ravens, as Jesus suggested, and instead we have lost our way within the physical world by forgetting that it is our natural home and the source of our well-being and sustenance. The ravens understand this dynamic, but we do not. Our avian cousins the ravens subsist in harmony with the plentiful gift of creation in a determinate and faithful pattern. God feeds them, and they live their lives in full expectation of this plentitude. By energizing the living food chains on which they are dependent, God cares for, supports, and feeds the ravens on a daily basis, as they well know. But many of us have stopped taking notice of the ravens as models of sustainable living within the overflowing largesse of God's bounty that we and the ravens rely on. Instead, we make war against this bounty through destruction of habitat and anthropogenic climate change—laying siege to the good world the divine provender has offered to us and the ravens alike.

The theme of God's attentiveness to animal well-being is again sounded in Jesus' teaching about the sparrows—a third bird species with which he especially identified. In biblical times, sparrows were cheap food—much like the now-extinct passenger pigeon, which I will discuss in Chapter 5, was in nineteenth-century America—and, according to the Gospel of Matthew, two sparrows could be sold for a penny for a quick bite to eat (10:29–31). In spite of the sparrow's seeming worthlessness, Jesus championed it as inestimably worthy of God's concern and compassion. We think of sparrows as negligible gray puffballs or maybe even flying vermin or extirpable pests that destroy crops or take over native birds' nesting places. Jesus knew that this is the attitude of many of us, in his time and ours, but counters that what we find to be useless rubbish in nature God embraces as worthy of care and compassion. "Are not two sparrows sold for a penny?" he asks. "And not one of them will fall to the ground without your father's will" (Matthew 10:29). God's eye is on the sparrow, as the early American hymn puts it. This, then, should be a comfort to all of us: as God bears witness to the suffering and death of the smallest creatures among us—even the little sparrows sold for a penny whom people in biblical times and today regard as most insignificant, even despicable—God will do the same for all other beings, including ourselves. As the old hymn rings out, "I sing because I'm happy, I sing because I'm free, for his eye is on the sparrow, and I know he watches me."

Indeed, birds of all sorts are everywhere in Jesus' life. But what religion scholars and believers alike have often missed is not only that birds were Jesus' constant friends but that God in Godself is identified *as a bird* by Jesus and the biblical authors, as I have argued here. The dovey bird-God who alighted on Jesus during his baptism ("When Jesus was baptized, the Holy Spirit descended upon him in bodily form as a dove"; Luke 3:21–22) harks back to the same birdy Spirit who brooded over the face of the deep in the book of Genesis ("In the beginning God created the heavens and the earth, and the Spirit of God moved over the waters"; 1:1–2) and who hovered over Mary at the moment of Jesus' conception in the Gospel of Luke ("The Holy Spirit will come upon you, the Most High will cover you"; 1:35). This avian Holy Spirit—this *animal God*, as I have suggested—was Jesus' regular companion, even as the many roosters, ravens, and sparrows he met along the way of his daily excursions provided him with guidance in his ministry to others. When he stood up in Nazareth in his hometown synagogue and said, "The Spirit is upon me" (Luke 4:18), he was invoking the presence of the very same Spirit-bird who created the world, miraculously made pregnant his mother, protectively flitted about him at his baptism, and instigated his life-changing journey into the wilderness. There was a time—and there still could be that time today—when God was a bird, because Jesus was intimate with, was reliant on, and was regularly accompanied by this winged divinity throughout his life.

## *Augustine and Natalist Wonder*

Beyond the Jesus tradition, it is not an overstatement to say that Christian thought about nature, and basically everything else, has been a long and extended dialogue with the early North African theologian and bishop Augustine of Hippo (354–430 CE). Augustine's writings are foundational to the self-understanding of the three main branches of worldwide Christianity: Roman Catholic, Eastern Orthodox, and Protestant. Augustine wrote on a wide range of theological topics, and no attempt to reconceive Christian faith in animist terms can proceed without an examination of his thought about God's all-pervasive grace and beauty within the natural world. Throughout his life, Augustine wrote consistently (some might say obsessively) about the Genesis creation story, using its hymn-like refrain, "And God saw everything that God had made, and behold, it was very good" (1:31), as a leitmotif for his own religious vision. Per Santmire, Augustine's theology is generally animated by the metaphor of fecundity. The goodness of creation and God's care and compassion for all beings, great and small, is the scriptural baseline that reverberates in all of Augustine's thought.[9]

In Augustine's poignant self-portrait *Confessions*, widely regarded as the first sustained autobiography in Western literature, he writes movingly of a series of religio-philosophical "conversions" he underwent prior to the full adoption of his mother's Roman Catholic faith at age thirty-one. Throughout his teens and twenties, Augustine alternately adopts as his own life orientation Cicero's paean to meditative philosophy called *Hor-*

*tensius*, the dualistic Gnostic writings of the then-powerful religious movement of Manichaeism, and the call to turn one's attention to a transcendent world of pure ideas in the Neoplatonic musings of the ancient Greek philosopher Plotinus. These three thought systems pave the way for Augustine's reading of the Bible and Christian witness as a record of God's benevolence within all things.

This theme is evident in *Confessions*, where, in the spirit of a sort of philosophical prayer, Augustine says to God, "I saw now, and was assured of it, that 'all you made is good,' and there is nothing in nature that you did not make. Though you did not make them all equally good, each is good insofar as it is made, and the entirety of them all is good, since God made them all 'good eminently.'"[10] In this passage, Augustine meditates on the proposition, as the Manicheans taught, that the universe consists of two substances, one good and one evil, or whether, as he eventually concludes, all things are good in and of themselves because God made all things as an extension of Godself. In this vein, Augustine comes to repudiate the Manicheans' denigration of earthy existence as depraved and corrupted.

Augustine makes this same point in his complementary magnum opus, *The City of God*. Augustine is not a pantheist, but, panentheistically, he argues that nothing is essentially contrary to God. God is all in all; everything is good because everything comes from God's supreme goodness; and nothing came into existence, nor is there any substantive order of things, that is structurally opposed to God:

> Accordingly we say that there is no unchangeable good but the one, true, blessed God; that the things which he made are indeed good because [they are made] from him. . . . Although, therefore, they are not the supreme good, for God is a greater good, yet those mutable things which can adhere to the immutable good, and so be blessed, are very good. . . . Consequently, to that nature which supremely is, and which created all else that exists, no nature is contrary save that which does not exist. For nonentity is the contrary of that which is. And thus there is no being contrary to God, the Supreme Being, and Author of all beings whatsoever.[11]

It may seem that fallen angels or evil itself are of a different essence than God; but divinity is everywhere, Augustine writes, and nothing is of a different origin from or of a different nature than God. Against the error that reality consists of two distinct natures or substances—one good and one evil, as the Manicheans said—Augustine, a theological *monist*, contends that the author of all things ensures the goodness of all things.

While Augustine takes issue with the Neoplatonists' disparagement of the world and the flesh as separate from God, he still at times writes under the sway of the metaphor of ascent. This influence is most apparent in Augustine's deployment of the category of "original sin." Paul says that "sin came into the world through one man [Adam], and death came through sin, and so death spread to all because all have sinned" (Romans 5:12). It is not clear what Paul means by this aboriginal transmission of sin to all things,[12] and nowhere does Paul use the phrase "original sin," but Augustine picks up this Pauline theme and proceeds to *biologize* it by specifying that through Adam—and, specifically, Adam's semen—the whole world becomes contaminated with sin: "Man, being of his own will corrupted and justly condemned, begot corrupted and condemned children. For we were all in that one man [Adam]. . . . For not yet was the particular form created and distributed to us, in which we as individuals were to live, but already the semen was there from which we were to be propagated; and this being degraded by sin, and shackled by the bond of death, and justly condemned, man could not be born of man in any other state."[13] As Elaine Pagels puts it, "semen itself, Augustine argues, already 'shackled by the bond of death,' transmits the damage incurred by sin. Hence, Augustine concludes, every human being ever conceived through semen already is born contaminated by sin."[14]

Augustine's pseudoembryological idea of original sin—humankind's hereditary tendency to wrongdoing springs from a "corrupt root"[15] of seminal fluid that virally transmits sin from one generation to the next—has, at times, burdened Christianity with the weight of an antisexual and antibody theology that stretches back to late antiquity. Nevertheless, while Augustine, at times, wrote negatively of fleshly *desire*—at one point in *Confessions*, he writes of "the time of my young manhood, when I burned to be engorged with vile things"[16]—he consistently valorizes flesh *itself* as the privileged site of God's disclosure of Godself as the Word of God in the person of Jesus. Again in the form of an address to God, he writes about the impact the Neoplatonists had on his thinking. But then he writes what happened when God brought to him a teacher "to acquaint [him] with certain books of the Platonists, translated into Latin from Greek": "What I found in reading them, not precisely in these words, but saying the same things in varied and very convincing ways, was this: 'At the origin was the Word, and the Word as in God's presence, and the Word was God. This was at the origin with God, and all things were made through it, and nothing was made without it.'"[17] Augustine concludes this passage with the keen insight that while the Neoplatonists rightly see that God, or the Word of God, is in some sense beyond this world, the Neoplatonists are wrong, thereby, to conclude that God is not a flesh-and-blood being. Why? *Because, according to Augustine, it is precisely in the bodily incarnation of God in Jesus that one rightly sees God for who God is.* "I did read there that the Word, God, is 'not born from flesh, or blood, or human desire, or of fleshly desire, but from God.' But I did not read there: 'The Word became flesh, to live with us.'"[18] The Neoplatonists were right to argue that God's Word was from God but wrong to locate God's Word—what they called "Reason" or "the *Logos*" and whom Christians later identified as Jesus—in an extramundane world of disembodied ideas. On the contrary, says Augustine, it is in *this* world, the world where "the Word became flesh, to live with us," to quote John 1:14, that God's Word in Jesus is *carnally* set forth. In reference to the notion of animism, Augustine sounds the theme of Jesus' fleshy body in order to articulate a model of God as always-already and everywhere present in the goodness and the bounty of the natural world. In this fundamental sense, Augustine saved Christianity from becoming a footnote to historical Neoplatonism—a religious philosophy of disincarnation—and put it on a path toward regarding this here-and-now corporeal world as the divinized medium through which God's robustly good and compassionate activity is supremely set forth. With this master stroke by Augustine, Christianity again became, as it always was, the once and future religion of embodiment.

Augustine's theology of the flesh makes special note of God's particular embodiment in the form of the brooding *mother bird* in the book of Genesis—a central biblical trope in my animist Christian proposal. In Chapter 1, I suggested that an avian thread ties together the depictions of the Holy Spirit in Genesis and the Gospels: in the first narrative, the feathered God of creation broods over the great egg of the world in loving embrace; in the second, this same divine being hovers over Jesus at the time of his baptism and the inauguration of his public ministry. In this vein, Augustine highlights the *Trinitarian* character of the bird-God's caring activity in Genesis. In one of his many commentaries on Genesis, he exegetes the opening verses that identify the Spirit as "brooding over the face of the waters" of creation (1:2). Augustine emphasizes the birdlike warmth and affection that the third member of the Godhead showers forth at the time of creation—an expression, he writes, of God's eternal commitment to the welfare of all creation:

> Above all, let us remember, as I have tried in many ways to show, that God does not work under the limits of time by motions of body and soul, as do men and angels, but by the eternal, unchangeable, and fixed exemplars of his coeternal Word and *by a kind of brooding action of his equally coeternal Holy Spirit*. . . . This action is not like that of a person who nurses swellings or wounds with the proper application of cold or hot water; but it is rather *like that of a bird that broods over its eggs*, the mother somehow helping in the development of her young by the warmth from her body, through an affection similar to that of love.[19]

Augustine's theology, in a word, is *natalist*. The distinctive action of God as Spirit is "like that of a bird that broods over its eggs," protectively tending to and bringing into full existence the nascent life-forms under her charge. Birdlike, the work of the Spirit is to lovingly give birth to and nurture others' well-being. I submit that Augustine's invocation of the winged God of creation who compassionately broods over all creation paves the way for a thoroughgoing animist reconfiguration of Christianity in the current setting. This reconfiguration lays bare the suggestive strands of divine animality present, but sometimes hidden, within the history of Christian thought. Harking back to Genesis, Augustine recovers the bird-God symbol of biblical-historical Christianity with razor-like clarity, opening the way forward for a continued retrieval of this symbol today.

Augustine's love of all of God's creation—brilliant celestial bodies in the sky, myriad creatures in water and on land, trees and flowers of infinite colors and textures, and the sensual pleasures of cool breezes, warm clothing, and daily nourishment—all of these many gifts he heralds as the expression of divine benevolence in everyday life. In recalling this abundance, his stirring rhetoric at the end of *The City of God* crescendos in a fever pitch of joy and praise:

> How can I tell of the rest of creation, with all its beauty and utility, which the divine goodness has given to man to please his eye and serve his purposes. . . . Shall I speak of the manifold and various loveliness of sky, and earth, and sea; of the plentiful supply and wonderful qualities of the light; of sun, moon, and stars; of the shade of trees; of the colors and perfume of flowers; of the multitude of birds, all differing in plumage and in song; of the variety of animals, of which the smallest in size are often the most wonderful—the works of ants and bees astonishing us more than the huge bodies of whales? Shall I speak of the sea, which itself is so grand a spectacle, when it arrays

itself as it were in vestures of various colors, now running through every shade of green, and again becoming purple or blue? Is it not delightful to look at it in storm, and experience the soothing complacency which it inspires, by suggesting that we ourselves are not tossed and shipwrecked? What shall I say of the numberless kinds of food to alleviate hunger, and the variety of seasonings to stimulate appetite which are scattered everywhere by nature, and for which we are not indebted to the art of cookery? How many natural appliances are there for preserving and restoring health! How grateful is the alternation of day and night! how pleasant the breezes that cool the air! how abundant the supply of clothing furnished us by trees and animals! Who can enumerate all the blessings we enjoy? If I were to attempt to detail and unfold only these few which I have indicated in the mass, such an enumeration would fill a volume.[20]

This passage is the keystone in the grand arch of Augustine's theology of nature. My reading of his oeuvre positions his analysis, in the *Literal Meaning of Genesis* commentary, of the particular work of God's avian Spirit in creation—the divine mother hen who gives birth to all beings—as a crucial moment within his soaring vision, articulated here toward the end of *The City of God*, of all things, at the beginning of time and into the present, joining together in a hymn of praise and thanksgiving for the unfolding beauty and wonder of the created order. Ultimately in Augustine, there is no ontological division between God, humankind, and all the other kinds that fly and swim and move above and within our common terrestrial home. God as the Word of God in Jesus' incarnation and as the brooding Holy Spirit at creation's birth is the God who creates and sustains the luminous marvel of the planets, the sweet fragrance of flowers, the colorful plumage of birdlife, the grand spectacle of the oceans, and the rich enjoyment of good food. We all come from and are made of God—a reality most beautifully symbolized by the embodied Jesus and the aerial mother bird, in Genesis and beyond, who hovers over all creation and its offspring.

## *Hildegard's Viriditas Pneumatology*

In the history of Christianity, animist sensibilities populate the medieval period as well, including the work of Hildegard of Bingen. Living on the Rhine River in central Germany, Hildegard was a twelfth-century monastic and mystical prophet who wrote Trinitarian theologies with special attention to the role of the Spirit in the world. At the age of eight years old, Hildegard's parents presented her to a religious order to prepare her for cloistered life. For the next thirty and more years, she became an anchoress and lived a quiet life as a recluse, walled into a monastic cell by the local bishop. But in midlife, Hildegard emerged from her cell to become a prolific writer, musician, artist, herbalist, abbess of her growing religious community, and even statesperson as she maintained influential relationships with bishops, kings, and emperors during the High Middle Ages. She was called the "Sibyl of the Rhine" for her wide-ranging impact on medieval culture through the power of her visionary writings.

In Hildegard's major theological work, titled *Scivias* (that is, in Latin, *Sci vias lucis*, "Know the ways of light"), she says she heard a voice from a living fire say to her, "O you who are wretched earth and, as a woman, untaught in all learning. . . . Cry out and relate and write these my mysteries that you see and hear in mystical visions. So do not be timid, but say those things you understand in the Spirit as I speak them through you."[21] Hildegard, being commanded by God to "cry out and write," becomes an oracle of the Holy Spirit. Though women were forbidden to exercise public leadership roles in the teaching ministry of the medieval church, the Spirit cut loose Hildegard's hesitant tongue and enjoined her to preach. Many of Hildegard's contemporaries, including many male clerics, saw her as filled with the Spirit and able to exercise the biblical role of prophet in a culture that needed her special message. And today, Hildegard, in her theological writings, musical compositions, and major medical works, is regarded by many people as a Spirit-inspired trailblazer for people who look for God's call in their lives as a subversion of male-dominated ecclesial and social orders.

The thrust of Hildegard's message is essentially ecological and biocentric, as we understand these terms today. Her pneumatology follows the arc of the biblical story, from the beginning of creation, when the bird-like Spirit brooded over the face of the deep, to the bird-God's hovering over Jesus at the time of his baptism, to the Spirit's outpouring of itself at Pentecost, when the founding disciples were crowned with tongues of Spirit-fire over their heads and spoke to the assembled gathering in different languages. Harking back to the earthen language of the Spirit in the biblical texts—the Spirit is breath, water, fire, and, in particular for the Rhineland mystic, a dove—Hildegard offers an avian model of the Spirit, much like Augustine's, in relation to the other two members of the Godhead. Two passages in particular from *Scivias* highlight Hildegard's theology of the winged Spirit:

> He who begets is the Father; he who is born is the Son; and he who in eager freshness proceeds from the Father and the Son, and sanctified the waters by moving over their face in the likeness of an innocent bird, and streamed with ardent heat over the apostles, is the Holy Spirit.[22]
>
> It is the Father who begot the Son before the ages; the Son through whom all things were made by the Father when creatures were created; and the Holy Spirit who, in the likeness of a dove, appeared at the baptism of the Son of God before the end of time.[23]

Hildegard's theology of the Spirit "in the likeness of an innocent bird" or "in the likeness of a dove" both signals her adoption of the historical Christian model of the Spirit in avian terms and dovetails with ancient and medieval ornithological understandings of how certain birds breed and nurture their young, much as the brooding Spirit-bird did at creation and Jesus' baptism.

Beyond Hildegard's theology of the dovey Spirit, she also wrote widely about a variety of other birds and their healing and medicinal properties—including the hawk, the peacock, the stork, the wren, and the cuckoo, among others. In her theological aviary, she is particularly interested in the nesting behavior of the pelican, whom she views as a symbol of Christ, in a manner similar to her vision of the dove, whom she sees as a symbol of the Holy Spirit. In this regard, Hildegard adopted the widely accepted medieval idea of the pelican as an infanticidal and self-sacrificial creature, who first slays her young hatchlings, on the one hand, and then, on the other, pecks at her own breast to bring up drops of blood in order to resuscitate her young. She writes, "When the pelican first sees her chicks hatch from her eggs, she thinks they are not related to her and kills them. When she sees that they do not move, she is sad and lacerates herself, resuscitating them with her blood."[24] It appears Hildegard was familiar with the *Physiologus* (Greek for "The Naturalist") or its many Latin interpretations and successors. The *Physiologus* was an anonymous second-century CE early-Christian bestiary that was popular and widespread during the Middle Ages. Ironically, in spite of its name, the *Physiologus* was less a study of animals and birds on the basis of early scientific observation and more a theological interpretation of the meaning of creatures on the basis of biblical and Christian symbolism. In this vein, the *Physiologus*, and its many medieval adaptations and expansions, states that it is well-known that the pelican chicks attack their mother at birth, causing the mother to kill the chicks, at which point, after three days, the mother pelican revives her young by pecking at and tearing open her own breast and pouring blood over the dead bodies of her children, bringing them back to life and thereby martyring herself in the process. The Christian allegory here is obvious: like the pelican who alternately, but justly, destroys and then self-sacrificially revives her young, so Jesus, upon the death of God's human children, saves and resurrects them by self-destructively shedding his own blood and giving up his own life on the cross—an act that is then reactualized in the bloody pelican-like sacrifice of the Eucharist.

The symbol of the pelican consumed with passion for her children is important to a variety of Western Christian writers from Dante and Thomas Aquinas to Shakespeare. Beginning in the medieval period and continuing into the present, the pelican noted in Hildegard is also powerfully set forth in European processional crosses, stained glass, and ecclesiastical seals.[25] In the modern period, the self-giving pelican with her chicks is featured in a bas-relief sculpture at the Princeton University chapel in Princeton, New Jersey, and in the vaulting bosses at Fourth Presbyterian Church in Chicago, Illinois. Two especially moving contemporary examples of Christian pelican art are the late-nineteenth-century ornate brass pelican lectern in Durham Cathedral, England, designed by George Gilbert Scott, and the pair of large sculpted pelicans that support a glass-tabled altar in the Cathedral of Saints Michael and Gudula in Brussels, Belgium. In the history of Western art, these images are referred to as "the pelican in her piety," with the mother pelican depicted as bending her long neck to her bosom as she lacerates her breast to revive and nourish her young. The message of the Eucharist is that the crucified Christ is the sanguinary food who feeds humankind with himself. Analogously, Hildegard's pelican spirituality, standing in a long line of pelican-based theology and art, articulates this message with self-sacrificial flesh-and-blood imagery that continues to resonate today.

Inhabiting a world of the divinized Spirit-bird (that is, the dove) and the many other birds who allegorized the story of Christ (for example, the pelican), Hildegard's Spirit theology is predicated on a vision of creation as a lush and fecund environment for growth and renewal. In creation, the Spirit reveals itself in ever-flowing streams and teeming oceans; luxuriant landscapes of wild forests and cultivated fields; the sweet-smelling air that enables and sustains the breath of all life; and, overall, in the green mantle that shrouds and protects the well-being of all creatures within the bountiful garden of the natural world. In the history of Christian thought, the Spirit is often relegated to the vague and colorless role of being a silent and passive intermediary between the other two members of the Trinity—the other two members who *really* count, namely, the Father and the Son.[26] But in Hildegard, the Holy Spirit is consubstantially, not merely nominally, coequal with the Father and the Son in purpose and activity: the Spirit is the dynamic life force who is bodied forth in the leafy green world of Earth community. In Hildegard, the Spirit is the power of nature's boundless fertility, the source of God's sustaining passion for the well-being of all things.

The theologian Elizabeth Dryer defines Hildegard's pneumatology in terms of "greening" (in Latin, *viriditas*) in order to describe, in rich color imagery, the Spirit's role in infusing the world of plants, animals, and humans with life-giving fecundity. Dryer writes,

> In addition to the Spirit's role as prophetic inspiration, Hildegard links the Holy Spirit with the term *viriditas* or "greening." She imagined the outpouring of the Spirit in natural rather than cultural metaphors. She combined images of planting, watering, and greening to speak of the presence of the Holy Spirit. Hildegard linked the flow of water on the crops with the love of God that renews the face of the earth, and by extension the souls of believers.
>
> *Viriditas* was a key concept that expressed and connected the bounty of God, the fertility of nature, and especially the presence of the Holy Spirit. . . . Hildegard describes the prelate who is filled with weariness (*taedium*) as lacking in *viriditas*, and counsels the neophyte in religious life to strive for "spiritual greenness."[27]

In Hildegard's *viriditas* theology, green is the dominant shade of color. At first glance, this may appear to be an odd choice because this color type seems distinctive of forested, riverine, and cultivated agricultural areas, not the vast swatches of landscape characterized by the more common earth hues of yellow and brown. If God's Spirit is everywhere in Hildegard, why does she define the Spirit's identity principally in grassy tones and ignore the colors we often associate with areas of sparse vegetation? While Hildegard's Rhineland watershed in the medieval period, as is the case today, was characterized by lush vineyards and a leafy forest canopy, I think her use of the term *viriditas* or greenness is less autobiographical and more expressive of her overarching vision of divine abundance as the unitive theme of Christian engagement with all things.

On the color wheel, green is a secondary color that combines the two primary colors of blue and yellow, and it stands opposite the color of red, the third primary hue in classical color theory. As derivative and combinatory of blue, the color of water, and yellow, a principal earth tone, green, chromatically speaking, is expressive not only of vegetation but also of bodies of water (as blended with the color blue) and earthen lands and countrysides (as blended with the color yellow and in juxtaposition with the color red). In other words, Hildegard's theology of *viriditas* or greenness is beautifully representative of both her major focus on God's luxuriant fertility and her wider interest in the other-than-green watery and terrestrial settings that make up the full expanse of verdant, divine providence.

Hildegard's green Spirit vision is metaphysically all encompassing, to be sure, but also practically medicinal. Hildegard writes that the green "firmament" that nurtures and protects planetary existence—the shelter of vivifying atmosphere and tree canopy that "covers" all things—is a living "garment" that testifies to God's endurance and compassion. In a vision, she sees, "The winds, air, and greenness of the earth, which are under the firmament of heaven, cover [humankind] as if they are a garment. This is because the flight and breadth of the winds and the sweet moisture of the air and the keen greenness of the trees and herbs . . . exhibit glory when they are fully obedient to God as he produces and maintains them."[28] In concrete terms, Hildegard's *viriditas* vision is the basis of her practice as a medicinal naturalist and botanical healer. Because Earth is a green mantle for the protection and healing of humankind and otherkind, it is a therapeutic source for everyday well-being. Undergirded by her knowledge of horticulture and herbalism, she developed in the twelfth century a well-renowned practice of plant-derived treatments rooted in her theology of the natural world as healing gift. Beginning with the Genesis story, in which primordial human being is created out of the earth's rich soil, Hildegard writes that creation is a balanced symbiosis between human need, on the one hand, and what she calls "vital energy" actualized through "beneficial herbs," on the other: "With earth was the human being created. All the elements served mankind and, sensing that man was alive, they busied themselves in aiding his life in every way. And man in turn occupied himself with them. The earth gave its vital energy, according to each person's race, nature, habits, and environment. Through the beneficial herbs, the earth brings forth the range of mankind's spiritual powers and distinguishes between them."[29]

Keeper of the soil and careful student of animal life, Hildegard published extensive catalogues of the curative properties of the flora and fauna she cultivated or observed in order to dispense a wide array of natural remedies to the ailing visitors at her monastery in Bingen. Like the ancient Greek text the *Physiologus* and its numerous early and medieval Latin extensions, Hildegard wrote *Physica* (Latin for "Natural science") as her own master inventory of the natural elements of her time—from plants, trees, rocks, and minerals to fish, birds, mammals, and reptiles—but now in order to analyze these elements' medicinal value along with their theological or symbolic significance. *Physica*'s culture-based medical anthropology is at its core grounded in a theory of spiritual and biological equilibrium. Each person consists of a balance of four qualities or "humors"—hot, cold, moist, and dry—wherein a person's health is guaranteed when this stability is maintained. Correspondingly, ill health results from one or more of these qualities falling out of balance. *Physica*, then, consists of an exhaustive record of the exact properties of each natural element and how these properties can be harnessed in the service of preserving the inherent balance of a person's qualities or humors.

As a practitioner of what we might call today "folk healing" or "vernacular medicine," Hildegard employed a wide variety of nature's gifts in her ministry of healing: the fruit of the beech tree to combat fever, heated white crystals to draw bad humors from the eyes, fresh catnip made into a poultice to heal ruptured lesions in the neck area, and the flesh of the house wren, cooked with water, lard, vinegar, and wine, to cure palsy. I see her role in *Physica* to be that of a traditional healer, or shaman, who channeled the vitalities of the natural world toward the ailing and hurting bodies and souls of the patients of her time. Much like the roles Moses and Jesus played in their own mediations of natural powers for healing purposes—remembering Moses' snake rite in Numbers and Jesus' mud-pie ritual in John—Hildegard as Christian shaman also deployed creation's vital powers to remediate disease and disability. Priscilla Troop, in her superb English translation of the Latin text of *Physica*, notes that the book was originally entitled by Hildegard herself as *The Subtleties of the Diverse Natures of Created Things*. This longer title makes more sense than the one-word Latinized title Hildegard's work came to be known by. God's good earth bears within itself a panoply of "subtle"—in the sense of being hidden or secreted—medicinal properties that the discerning practitioner can release through her healing arts. In the manner of Heideggerian *techne*, the Hildegardian healer seeks to actualize the biodynamic power of nature's latent energies.

Hildegard's medical practice—rooted in earth's *viriditas* for all beings' renewal—is beautifully set forth in one particular incantatory ritual in *Physica*. Hildegard calls this her "Greenish Earth" (in Latin, *terra subviridus*) ritual as she brings together both her creation-centered Trinitarian theology and her shamanistic practice of using quantities of dirt, carefully placed around the patient's body, to heal victims of numbness. This practice stems from Hildegard's analysis of four different types of dirt according to each type's (1) color (white, black, red, and green), (2) corresponding medicinal properties, and (3) relationship to her overarching quartet of natural humors (hot, cold, moist, and dry).[30] The white, black, and red types of earth are not medically potent but crucial for food production, while the green type is not agriculturally productive but very efficacious therapeutically. White earth, while neither primarily hot nor cold, is characteristically dry but still able to retain large amounts of moisture, making it ideal for fruit trees and vineyards. Black earth, which typically is cold and moist, is generally well balanced vis-à-vis all four humors and therefore is ideal for fruit trees as well as other modes of food production. Red earth is characterized by a natural equilibrium of moisture and dryness, though its hot and cold ratio is unclear; it initially sustains fruitful agriculture, though many of the crops it does support do not come to fruition. And green earth, the type of soil used in Hildegard's numbness ritual, is noted for its hot and cold evenness, and while it is not balanced between moisture and dryness and therefore not good for farming, green earth is ideal as a natural medicinal. As such, Hildegard, in a manner resonant with her *viriditas* theology, prescribes the following green-earth regimen for people who suffer from numbness:

> If someone is overwhelmed by numbness, another person should take a bit of the [greenish] earth from the right and left side of the bed where the sick person's head is, and in the same way take earth from near the person's right and left foot. While he is digging it he should say, "you, earth, are sleeping in this person, N." And he should place the earth which had been taken from both sides of the patient's head under his head, until it grows warm there. In a similar manner, he should place the other earth under his feet, so that it might receive heat from them. When the earth is placed under his head and feet, this should be said, "You, earth, grow and be useful in this person, N., so that he may receive your vital greenness, in the name of the Father, and the Son, and the Holy Spirit, who is the all-powerful, living God." This should be done for three days.[31]

Hildegard's three-day ritual consists of taking handfuls of green soil and packing it around the head and feet of a person suffering from numbness, perhaps a neuropathic disorder characterized by paralysis, pain, or tingling sensations. The goal is for the earth itself, thoughtfully positioned near the patient, to be its own agent of healing and to radiate its restoring warmth into the numbed body of the sufferer. Reminiscent of the Genesis story wherein human beings are said to consist of mud and clay, the ritual presupposes a perfect isomorphism between the sick person's earth-based constitution, on the one hand, and the soil itself, on the other, as noted in the healer's incantation, "You, earth, are sleeping in this person, N." After this pronouncement, the healer concludes the ritual and intones, "You, earth, grow and be useful in this person, N., so that he may receive your vital greenness."

While not referenced in the text, Hildegard's green-earth ceremony appears almost identical to the Johannine Jesus' mud-pie performance. In both instances, earth is sacramentally celebrated for its healing powers in a step-by-step exercise wherein soil is packed onto or nearby the body of suffering individuals who benefit from the curative energies of creation. *In sum, earth's animist fecundity is the common green thread that ties together both the Jesus and the Hildegard healing narratives.* In the one case, Jesus uses "anointing clay" to heal the blind man; in the other, Hildegard enjoins earth's "vital greenness" to heal the patient with numbness. Placing the two rituals together side by side, the echo across the centuries is resounding. In John 9, Jesus the earth-shaman's powerful act of renewal and transformation now provides the animating context for Hildegard's lifelong ministry of green-earth healing as set forth in her compelling *viriditas* ritual.

## *Rewilding Christian Worship*

In Chapter 4, I hope to round out my portrait of historical Christian animism with a full analysis of John Muir's theology of creation. Moving from the ancient and medieval periods in this chapter to the modern period in the next provides a thoroughgoing chronology of the development of sacred earth theology.

But in this conclusion to this chapter, I ask a final question about the practice of religion rooted in divine enfleshment. This question centers on affective Christian worship, the heartfelt habit of people of faith, seeking to live lives of reverence and gratitude before God, as they practice the presence of God in their private and congregational settings. Specifically, in the light of this book's principal theme, my question focuses on the worship of God in the registry of divine animality. My question is whether the basic grammar of Christian belief can include the practical devotion to God's full embodiment within the plethora of creation, from animals and plants to landscapes, bodies of water, and the atmosphere above. I am asking, in orthodox Christian belief, whether it is right and proper to reimagine worship today as a joyous exercise of love and adulation of God as God incarnates Godself within the created order of all things. Or to put it another way, is it permissible to worship divinized nature in inner-Christian terms? And if this is so, what would devotional ritual look like that appeals to God and the world together as the vital source, and common referents, for praise, hymns, chorus, sermons, sacraments, offerings, and prayers? How might Christian worship be practiced today that worships God and nature together as one in unity, in purpose, in essence, and in love?

My answer to these questions is personal and anecdotal and begins with a historical preamble. In the 1940s, the country of Costa Rica was riven by civil war. In a small country, thousands died in bloody internecine battles marked by assassinations, street violence, voter fraud, and the collapse of economic development and civil society in general. In 1948, José Figueres Ferrer, a rebel leader and then president of the divided country, made the radical proposal that the best way to end the fighting would be to abolish Costa Rica's military. A visionary leader of the time, Figueres could see no end to the protracted conflicts, and so he offered the solution of complete demilitarization as the best means for establishing the safety and well-being of the people of Costa Rica. Now enshrined in the country's new constitution, Costa Rica's elimination of its military began a long-term democratic experiment that has ensured the protection of its natural environment, widespread economic development, guaranteed access to education and health care—and, unlike most other Central American countries during the past seventy years, peace and security for its citizenry.[32]

At around the same time as Figueres's democratic experiment, some three thousand miles away, another, less notable development was taking place in a rural farming village called Fairhope outside of Mobile, Alabama. There an intentional group of young men and women in the Religious Society of Friends, or Quakers, had been gathering for prayer and discussion about how best to respond to the compulsory military draft required of all young adult American males in the immediate aftermath of World War II. In general, Quakers practice a "peace testimony" in which they forswear any direct personal or corporate involvement in armed conflict to settle disputes. In the same year, 1948, that Figures persuaded Costa Rica's constitutional assembly to abolish its military, four of the young Quaker men in Fairhope made public to the authorities their decision not to register for the Selective Service draft. The men could have registered for the draft and simultaneously claimed conscientious objector status but instead decided to make a point concerning their opposition to the state bureaucracy of war by publicly declaiming against military registration during peacetime. In response, the Quaker dissidents were arrested, and in 1948, they were sentenced by a U.S. district judge to serve one year and one day in a minimum-security prison. They were released after serving four months and a day.[33]

Now realizing that the United States was not a safe place to live and raise a family if they wished to remain free of involvement in military service, the four young Quakers, their families, and other members of the Fairhope Meeting began to consider emigration from their homeland. To explore potential locations, several advance parties were sent out to scout new locations for the community in the Americas. While some families ended up moving to Canada, the majority of the soon-to-be expatriates decided on Costa Rica and uprooted themselves in 1950 to tropical forest land they had purchased in the north-central part of the country in a small hamlet called Monteverde, or Green Mountain. The choice of Costa Rica as a safe haven free from compulsory conscription was largely determined by Figueres's disbanding of the Costa Rican armed forces in 1948. Thus, the Monteverde Quaker Colony began in 1951. While the Quaker community initially focused on dairy farming as a primary means of income, it soon became an environmental leader in Costa Rica, and eventually globally, as it began to set aside, preserve, and study tens of thousands of acres of primeval cloud forest in confederation with government and conservation allies. Today Monteverde is rightly celebrated as a green pioneer in the study and preservation of biodiversity in the tropics.[34] As one of the original founders of Quaker Monteverde, Wilford ("Wolf") Guindon, said to me when I asked him why he and the founders shifted from dairy production to biological conservation, he replied simply, "I fell in love with the trees."[35]

Over time, the Friends community in Monteverde grew in size from a handful of original founding families to a sizeable town of Quakers, biologists, small-business people, farmers, and educators, among others. In ensuing years, commerce diversified beyond the early dairy farming; roads and infrastructure were built along with supermarkets, banks, clinics, and schools; and today, Monteverde hosts 250,000 visitors annually, from adventure seekers to environmental studies academics such as myself, who visited the area in 2004 with my family during a yearlong sabbatical. Subsequent small waves of North American and European immigration and intermarriage between the children and grandchildren of the founders and local Costa Ricans has ensured Monteverde's demographic vitality and positioned the town for success in the years to come. But amid the ever-growing whirl of changes and activity, the one abiding spiritual constant in the community has been regular worship in the local Quaker meetinghouse—which, today, is a stunning wood-framed and glass-enclosed "church" perched on a small rise deep in the woods of the cloud forest.[36]

Housed within the Monteverde Quaker school complex and library, the meetinghouse has undergone four major changes since the colony's founding in 1951. The first meetinghouse was a small wooden building, a "disused squatter shack," as the builder and author Shannon McIntyre puts it.[37] This modest space was initially repurposed as a community room, in the early 1950s, after which a Quaker elementary school was built, and part of this school became for a while the Sunday-morning Quaker meetinghouse. Then, in 1957, a large meetinghouse with pews, windows, and a movable staging area was constructed as an attachment to the growing school complex. This erstwhile meetinghouse still stands, but in 2013, the Quakers joined forces with local townspeople and visiting builders to erect, in the tradition of Amish community barn raising, a beautiful new timber-framed meetinghouse made of rich reddish-brown cypress and large plate-glass windows that open out to all the sights and sounds of the tropical forest. This 2013 project brought together experienced woodworkers and scores of volunteers using hand tools and wooden pegs to fashion together the new Quaker meeting for worship.

On a recent trip back to Monteverde, I enjoyed the meeting for worship on a cool autumn morning in the new meetinghouse. Here I actively *felt* the full impact of the ancient biblical truth that all nature is sacred because God has poured out Godself into all things—trees, plants, rivers, birds, reptiles, mammals, insects, and rocks—and saturated all things with divinity. Monteverde Friends practice a form of "unprogrammed," or leaderless, Quakerism, in which the congregation sits in silence awaiting the movement of the Spirit to stir one of the members to stand up and speak, an event that sometimes does and sometimes does not happen. On the morning Audrey and I attended the service, we sat in silence for about half an hour, and then various individuals spoke up on a variety of topics in a bilingual Spanish-English environment for the second half hour. During the worship time, some quoted and commented on the scriptures, others sang brief songs of wonder and adulation, while others stood to sway back and forth and raise their hands in gestures of worship and praise.

In this mingling of Spanish and English voices, spoken and sung, and nestled within this built setting of forest-hewn pews, generously sized open windows, and a cathedral-like timbered ceiling, I encountered the jungle outside and the church community inside as one and the same reality. Unlike many Christian churches that intentionally separate the built structure from the natural world, this Quaker mountain sanctuary self-consciously invited the local flora and fauna into the meeting space. I heard streaming through the open windows, in the silence and the mix of voices, the bell-like bonking call of three-wattled bellbirds, a white-and-chestnut-brown-colored bird of the neotropical rainforest; the steady, thunderous growling of howler monkeys, large dark-haired primates who roar like lions and roam high in the tree canopy; and the sweet perfume and pitter-patter of raindrops, on leaves, grass, flowers, and the meetinghouse roof itself, where high winds, misty clouds, and steady rain are almost everyday occurrences. Just outside the meetinghouse were massive strangler fig trees, myriads of hummingbird species flitting about, and brightly colored mosses and orchids liming the green woods.

But what I remember most in this Quaker assembly was the eerie, flute-like song of black-faced solitaires, little gray, black-masked, and orange-beaked birds, whose calls echoed throughout the lofty forest surrounding the meetinghouse. I was mesmerized by these small birds' hollow-sounding but hauntingly pleasant musical notes. Undeniably, I felt as if I were back in Pennsylvania. The moving, echoey melody of the black-faced solitaire reminded me of my previous spellbound encounters with the song of the wood thrush during languid afternoons in the Crum Woods. I knew the Central American black-faced solitaire and the North American wood thrush were kindred birds, both thrushes, each of whom sings its own high-pitched, liquid, flutey harmonies of rising and falling notes. The Quaker meetinghouse gave shape and definition to the many melodies and sights that came to me that morning. In turn, the play of sound, vision, and fragrance invested my worship practice with a rich nuance and sensual quality generally unfelt by me in more conventional religious settings.

In the Costa Rican cloud forest, I experienced God's Spirit at play in the lush foliage circling the meetinghouse. I came face-to-face with pneumatology gone wild—the magic of feral Spirit flying through the trees, cascading in the waterfalls, and treading on the ground of this high mountain wilderness. Indeed, I felt that worshipping God and God's creation *simultaneously* were not opposing movements but cooperative gestures that mutually support and strengthen each other. I remembered Paul's saying, "Eye has not seen, nor ear heard, nor entered into the human heart, the things God has prepared for those who love God" (I Corinthians 2:9), and I encountered the literal, viscous impact of this scripture coursing through the veins of my heart and the depths of my consciousness. In worshipping God and nature together, in worshipping God-*as*-nature, I felt the rapturous truth of the ancient teaching that the Word has now become flesh and is dwelling among us. Through silence and many voices (human, avian, and vegetal voices alike), I touched and was touched by the lush green God—the God of Jesus, Augustine, Hildegard, and the Quakers—of historical Christian witness.

CHAPTER 4

# "Come Suck Sequoia and Be Saved"

## *John Muir's Christianimism*

In Chapter 3, I traced the growth of sacred nature theology from the ancient period in the Jesus tradition, the late antique period in Augustine, and the medieval period in Hildegard of Bingen. In this chapter, my aim is to complete the arc of this historical portrait with an analysis of John Muir's ecstatic nature religion as quintessentially expressive of the subtle dialectic between Christianity and animism at the center of this book's thesis. In the modern period, I turn to Muir as a model for a thoughtful and nuanced celebration of the full enfleshment of God's Spirit within all things: songbirds and waterfalls, mountain streams and alpine meadows, Sequoia trees and Douglas squirrels. Many contemporary scholars of Muir (and, in Muir's time, his own father) have failed to understand the scriptural-animist coherence of his project, what we might call *Christianimism*. It is the integral unity of Muir's divinized and verdant vision of the world that stands out in a reading of his voluminous works. Muir's original entwining of the ideas of divine subscendence and nature's splendor creates an eco-theological tapestry of profound beauty that continues to inspire awe and wonder today.

John Muir was a nineteenth-century/early-twentieth-century farmer, inventor, explorer, political activist, amateur geologist, nature writer, and religious visionary. Born in Scotland in 1838 to a pious family and erstwhile Calvinist father, he immigrated with his family to a Wisconsin farmstead as a young boy. Later, he matriculated at the University of Wisconsin and completed an array of courses in math, chemistry, geology, botany, and the classics. Shy of completing his university degree, Muir began a lifelong wanderlust of the Florida Everglades, the Colorado Grand Canyon, and California's Sierra Nevada mountain range. On foot, he traveled thousands of miles across the North American landscape. He marveled at the fragrance of cedar in the high mountain air, the beautiful melodies of riverine songbirds in the vernal streams of Sequoia meadows, the invigorating spray of great waterfalls through the canyons of Yosemite, and the glistening polish of glaciated valleys most recently formed during the last great Ice Age of one to two million years ago.[1]

As a prophetic voice for political change, Muir cofounded the Sierra Club and lobbied local and national politicians to save his beloved Northern California settings of Yosemite, the Hetch Hetchy Valley, and what became known as Sequoia National Park. He won two out of the three of these battles, but, sadly, he lost the struggle to preserve Hetch Hetchy from being flooded in service of San Francisco's growing municipal water needs. For Muir, it was the pending conversion of Hetch Hetchy into a damned reservoir that inspired some of his most passionate diatribes. In 1903, Muir journeyed with President Theodore Roosevelt on a three-day camping trip into the Yosemite Valley and mountains to experience primeval wilderness unfettered by corruption, politics, and bureaucracy. He made the case for a new vision of North American landscapes as inestimably worthwhile for their own sake, not as developable *resources* but as public lands that are inherent and wondrous *goods* unto themselves. "Any fool can destroy trees. . . . It took more that three thousand years to make some of the trees in these Western woods,—trees that are still standing in perfect strength and beauty, waving and singing in the mighty forests of the Sierra. Through all the wonderful, eventful centuries since Christ's time—and long before that—God has cared for these trees, saved them from drought, disease, avalanches, and a thousand straining, leveling tempests and floods; but he cannot save them from fools,—only Uncle Sam can do that." "These mighty forests of the Sierra" that "God has cared for" are worthy of preservation, not because of their utilitarian or financial importance, Muir writes, but because of the invaluable role they play as vibrant communities of complexity and beauty that inspire a sense of reverence and grandeur among all who visit their environs.[2] By 1906, Roosevelt, citing his memorable Yosemite expedition with Muir, conserved 230 million acres as nature reserves to be maintained in perpetuity as biological refuges and natural wonders.

Muir was the primary driver of the fin de siècle environmental crusade to set aside vast tracts of pristine wilderness as inspirational sites for uplifting the human spirit. Nature's value resides not only in the ecosystem services it performs—water filtration, erosion prevention, nitrogen fixation, plant pollination, and the like—but, moreover, in its capacity to serve as a sacred refuge for renewal and inspiration in the lives of the people who care for and enjoy its many gifts and splendors. Muir was the green prophet of the natural world as God's holy temple in which the aesthetic and religious regeneration of people's deepest selves is accomplished daily. It is primarily for this reason—the renewal of the soul—that the cathedral glory of the natural world must be saved. Harking back to the preservation of Yosemite National Park and looking ahead with foreboding at the likely damning of the Hetch Hetchy Valley, Muir wrote, "Everybody needs beauty as well as bread, places to play in and pray in, where Nature may heal and cheer and give strength to body and soul alike."[3] For Muir, wild nature is a joyous wonderland of unplanned delights—rare and exquisite species of plants and animals, bracing thousand-foot mountain waterfalls, vistas of astonishing beauty unmatched by any human-built structures—that cries out for honor and protection. In a world saturated with divinity, scenic but fragile wonders abound, and it is the duty of nature's guardians to steward these treasures and ensure the wellness, healing, and replenishment of mind, body, and soul for future generations.

## *Indian Removal in Yosemite*

Nevertheless, Muir's wilderness evangelism came at a price. In order to render sites such as Yosemite public trusts, America's eco-guardians decided that the original inhabitants who had faithfully cared for these lands for millennia had to be assimilated to American society or driven off their home ground, in order to make room for nature tourists to enjoy the Sierra Nevada's scenic marvels. Muir, in particular, and the national parks movement, in general, have been rightly criticized for cordoning off public lands and denying to Native peoples historically settled there the right to enjoy their homes and traditional hunting and fishing economies. At the height of the industrial revolution, Muir's rhetorical brio advanced the novel idea that untrammeled wilderness sites such as Yosemite are a spiritual commonwealth that should be preserved for posterity. But Yosemite and other eventual parklands, shaped by aboriginal human impacts for thousands of years, were not uninhabited, pristine landscapes. Rather, they were models of how human and more-than-human communities had arduously but amicably coexisted since the first Ice Age, when humans initially came to North America.[4]

I have not found references in Muir to endorsing policies of forced removal of California's Indians to places outside of what is today Yosemite National Park. But in general, Muir appears deeply conflicted in his attitudes to his Indigenous Sierra coinhabitants, the Ahwahneeche, a cultural-linguistic community who were part of the larger California Miwok people.[5] On the one hand, he praises their sustainable existence within the High Sierra ecosystem:

> Indians walk softly and hurt the landscape hardly more than the birds and squirrels, and their brush and bark huts last hardly longer than those of wood rats, while their more enduring monuments, excepting those wrought on the forests by the fires they made to improve their hunting grounds, vanish in a few centuries. How different are most of those of the white man, especially on the lower gold region—roads blasted in the solid rock, wild streams dammed and tamed and turned out of their channels and led along the side of canyons and valleys to work in mines like slaves.[6]

For Muir, white industrialists and colonists violently blasted their way into Yosemite—most likely a reference to the California Gold Rush of 1849 and the concomitant Mariposa War of 1851[7]—while the Yosemite Indians hunted and fished and built their homes simpatico with a well-maintained ecosystem. Like Muir, the Ahwahneeche also experienced Yosemite as a sacred landscape in balance with its human and other forestal neighbors, a "special place the Creator had filled with all they would need, including trout, sweet clover, potent medicinal plants, roots, acorns, pine nuts, fruits, and berries in abundance, as well as deer and other animals."[8]

But while Muir lauds his Indigenous neighbors for respectfully maintaining the integrity of this bounty over and against Anglo settlers, he also criticizes them, ironically, for being no different from their nonnative counterparts as a people who are not "natural" and, of all things, "unclean":

"Most Indians I have seen are not a whit more natural in their lives than we civilized white. Perhaps if I knew them better I should like them better. The worst thing about them is their uncleanliness. Down on the shore of Mono Lake I saw a number of their flimsy huts on the banks of the streams that dash swiftly into that dead sea,—mere brush tents where they lie and eat at their ease."[9] It is bigoted comments such as this that helped to fuel the expulsion of first peoples from their original territories and deprived them of their sacred sites and ancestral homelands. In my judgment, Muir's thought is crucial for articulating a romantic-animist vision of the world that is biblically resonant. But at what price? Jedediah Purdy asks, "Muir and his followers are remembered because their respect for non-human life and wild places expanded the boundaries of moral concern. What does it mean that they cared more about 'animal people' than about some human beings?"[10] While not explicitly lobbying for Native peoples' exclusion from Yosemite, Muir's noxious rhetoric about the unnatural and unclean Indian paved the way for an eco-fascist push to dispossess California's Indigenous populations of their rights to peaceable living and self-determination. The national parks system that Muir helped to found—"America's best idea," in the words of Ken Burns's eponymous 2009 PBS series—emerged as a racist land-grab ideology predicated on the assumption that Native presence contradicts the conservation ideal of untamed wilderness. "America's best idea" was also an extension of America's ongoing exercise in Indian removal, assimilation, and extermination. As Carolyn Merchant writes, "National parks and wilderness areas were set aside for the benefit of white American tourists. By redefining wilderness as the polar opposite of civilization, wilderness in its ideal form could be viewed as free of people."[11] This is the ugly irony of America's white conservation idyll of a depeopled wilderness: it invites non-Native people into its leafy environs for renewal and recreation while at the same time ethnically cleansing a carefully preserved biosphere of its original caretakers.

Muir's heartrending disregard of Native communities in his efforts to protect America's natural wonders is a growing stain on his legacy and the movement he nurtured and led. Today, this movement remains an important, if fundamentally defective, first step toward reimagining the natural world, not as exploitable resource but as intrinsically valuable and deserving of thoughtful preservation.[12] In the Victorian era of land barons and wealthy industrialists, Muir, as a leading writer, publicist, and provocateur, deployed ethical, religious, and scientific ideas of loving and saving the undeveloped countryside as a sacred commons for personal enjoyment and sustenance. I regard Muir, warts and all, as the profoundly flawed patron saint of the American environmental movement. In Yosemite and similarly wondrous places, Muir's vision of a divinized world is desperately needed today—even though this vision is tainted by a racist imaginary of depopulated wilderness that has had, and continues to have, toxic reverberations that impact first peoples' cultural heritage, natural rights, and future well-being.[13]

## *The Great Code*

In *The Great Code: The Bible and Literature*, the humanities scholar Northrop Frye argues that the Bible generates the "mythological universe" within which Western civilization has operated. He writes that all fields of human thought and endeavor in the West have been shaped by the Bible. Read as a literary whole, the Bible provides the root metaphors and narrative structure for conceptualizing the meaning and truth of all facets of reality. From physics, cosmology, and psychology to literature, music, and art, the Bible functions as the orienting cognitive framework for understanding space and time, the purpose of biological existence, the meaning of the inner life, the importance of family, the role of violence in forming culture, the nature of morality, and the shape of things to come.

Frye does not argue that this "great code" is scientifically sound, historically accurate, ethically praiseworthy, or even religiously true. His case for the centrality of the Bible in history and the present day is descriptive and analytical, not advocatory and evangelistic. "Clearly," he writes, "the Bible is a violently partisan book: as with any other form of propaganda, what is true is what the writer thinks ought to be true; and the sense of urgency in the writing comes out much more freely by not being hampered by the clutter of what may actually have occurred."[14] The Bible's dramatic narration of the sweep of human history is the measure of its hold on the Western imagination—notwithstanding the disputed status of its scientific and historical truth claims, on the one hand, and the dubious character of some its moral and religious teachings, on the other. The point of the Bible, and its staying power through the millennia, is its pitch-perfect capacity to describe with poetic potency the significance of everyday events vis-à-vis a divinity or life force that animates, sustains, and governs all things. It is the Bible's *figuration* of the world as enlivened by unseen forces that renders its many complicated and contradictory story lines endlessly compelling. In Western cultures, the Bible's lyrical vision of nature, society, and humankind's role therein continues to provide the founding optics through which many people have understood and still understand themselves today in communion and contestation with life's higher power and daily opportunities and obstacles.

"Father made me learn so many Bible verses every day that by the time I was eleven years old, I had [memorized] about three-fourths of the Old Testament and all of the New by heart and by sore flesh. I could recite the New Testament from the beginning of Matthew to the end of Revelation without a single stop."[15] Growing up along the southeast Scottish coastline in the small village of Dunbar and then moving to rural central Wisconsin, Muir was encouraged—and, at times, forced by the threat of physical discipline—to memorize every verse of the Bible. He accomplished this feat (or nearly so) at an early age and in his mature years patiently and seamlessly wove biblical language and imagery into the rhetorical fabric of his ecological vision. Of course, no child should be compelled to memorize anything under the threat of corporal punishment. But in spite of this traumatic beginning, Muir made the Bible's "great code" his own and later deployed its imagery in his nature writing with incredible originality, skill, and vigor. Muir was a consummate storyteller, skilled essayist, practiced letter writer, and gifted sermonizer. In all of these different genres, he threaded his arguments and observations with direct quotations from, and subtle paraphrases of, the biblical texts in a manner that is rich with aesthetic beauty and religious fervor. This practice of interlacing scriptural imagery and nature observation is especially characteristic of Muir in his thirties and forties—a period at the height of his literary output and political activity in the 1870s and 1880s. But this practice characterizes his body of work in general, even in the last years of his life, as I will show here.

## *The Water Ouzel*

Muir throughout his life recorded copious journal entries and wrote numerous letters to family and friends.[16] These initially unpublished writings contain the basis of many of his later antemortem environmental books, including *The Mountains of California* (1894), *My First Summer in the Sierra* (1911), and *The Story of My Boyhood and Youth* (1913). As well, there are numerous posthumous collections of Muir's journals, letters, and other occasional writings, including *The Life and Letters of John Muir* (1923), *John of the Mountains: The Unpublished Journals of John Muir* (1938), *John Muir: His Life and Letters and Other Writings* (1996), and *John Muir: Nature Writings* (1997). Muir's major themes—creation is God's first temple, human happiness depends on the well-being of all creatures, and salvation is grateful subsistence within a Spirit-filled universe—are articulated in a consistent rhythmic pattern in these principal books and essay collections.

The scriptural tenor—the great code—of this body of work is foregrounded at the outset. Muir writes his life in sonorous biblical cadence. Along with the word *God*, he regularly capitalizes the words *Nature*, *Beauty*, and *Spirit*, signaling to readers his understanding of the functional equivalence between the biblical deity, on the one hand, and corporeal divinity within the environs of Yosemite and the like, on the other.[17] At nearly every juncture of Muir's many observations, he sounds the depths of scriptural imagery to describe the wonders and intimacies of nature's warm heart.

One stirring narrative, in which Muir seamlessly integrates the natural and the sacred, takes place in southeastern Alaska on a canoe trip along a dangerous inland channel. In February 1880, in a letter about this fateful trip to a reader of one his earlier nature essays, he writes of a comforting visit from a chunky, gray musical bird called the water ouzel (now known as the American dipper) amid glaciers and icy fiords threatening to close in on his boating expedition. The ouzel, uniquely, is North America's only aquatic songbird. It sings and frolics along clear inland mountain streams and swims underwater to catch its food. Muir often writes about this playful songster and admires its plucky spirit diving for a meal in the midst of pounding cascades and rushing rivers. I think the ouzel reminds Muir of himself: steadfast, fearless, persistent, optimistic. He especially likes how the water ouzel sings and sings even when its voice cannot be heard over thunderous waterfalls. In the Alaska letter, Muir recalls the flight of one particular hearty ouzel out to his imperiled canoe party—he describes it as a divine visitation—and imagines the little bird speaking comfortingly to him and his fellow voyagers, as if to say, "Fear not, for I am with you":

> I'm glad you like my wee dear ouzel. His is one of the most complete of God's small darlings. I found him in Alaska a month or two ago . . . [but] fearing that we would be frozen in for the winter, and while pushing our canoe through the bergs, admiring and fearing the grand beauty of the icy wilderness, my blessed favorite came out from the shore to see me, flew once round the boat, gave one cheery note of welcome, while seeming to say, "You need not fear this ice and frost, for you see I am here," then flew back to the shore and alighted on the edge of a big white berg, not so far away but that I could see him doing his happy manners.[18]

Muir's recollection of "one of the most complete of God's small darlings," the water ouzel, and its heartening proclamation, "You need not fear this ice and frost, for you see I am here," is reminiscent of God's statement to the prophet Isaiah, "Do not fear, for I am with you" (Isaiah 41:10). More to the point, I think, Muir's ouzel encounter recalls Jesus' reassuring words to his disciples in their own dangerous maritime adventure, when Jesus walks out onto the storm-tossed waters of the Sea of Galilee and tells his terrified devotees, "It is I, do not fear" (Mark 6:50). As the biblical God earnestly strives to give comfort to the afflicted, including Jesus' followers at sea, so the sacred bird of Muir's scriptural imagination seeks to bring hope to Muir and his anxious seafarers. Muir translates Jesus' words of encouragement to his disciples into an imagined sign of avian hope: the water ouzel, normally a landward bird, takes leave of his typical habitat and flies out to Muir's scared party to reassure them that the shoreline is within reach. In Muir's biblical retrievals and his experience of divinized nature, God is present everywhere, saying, *Do not be afraid, do not lose hope, because I am with you to comfort and to guide you.*

In the water ouzel letter, Muir's subtle use of the Gospel story of Jesus on the water is typical of his overall biblical eco-hermeneutic. His frequent cross-referencing between nature observation and scriptural citation is not an exercise in Victorian-era rhetorical extravagance or extraneous biblical proof-texting. Biblical citation for Muir is neither ornamental nor irrelevant to the main thrust of his discourse. In fact, it is the warp and woof of that discourse itself. In addition to salting his writing with scriptural passages and allusions, Muir, as in the water ouzel example, often concludes a short story or essay with a particularly poignant biblical citation, as if to say, *The point of my exposition, and the point of the Bible, is the same.* The Bible illuminates Muir's intimacy with nature, and in turn this intimacy is deepened by Muir's constant reference to the biblical celebration of creation. *For Muir, to inhabit the world of the Bible, and to be a denizen of the natural world, is to be a citizen of one and the same universe of meaning.*

My study of Muir further notes that his deep reading of the Bible gives way to a prioritization of a particular "working canon," so to speak, that he uses to guide his interweaving of sacred texts and sacred nature. Indeed, interpreting the entirety of Muir's oeuvre reveals how he uses the *full sweep of Jesus' life span* in the Christian Gospels as the basic prototype for his own immersion in the wonders of the natural world. Analogically, many of the most significant events in Jesus' life—going to temple, baptism, departure into wilderness, Sermon on the Mount, Lord's prayer, driving out temple money changers, evangelistic mission, farewell discourse to disciples, Eucharist, and crucifixion—have a resonant trace element in Muir's autobiography as well. By studying the shape of Jesus' life pattern in Muir's liberal quotations and paraphrases of Christian scripture, one can discern the core structure that animates the themes and plotlines within his literary art. The key to understanding Muir's mission to love and save wilderness, I argue here, is to make sense of the biblical "great code" that informs all aspects of his life and thought.

## *The Two Books*

One entry point into understanding the scriptural contours of Muir's project is through his initial writings on the role of the church and customary ritual practices, such as baptism, within traditional Christianity. In some of his earliest extant letters that were written to his younger brother, David, Muir profoundly rethinks the meaning of church according to his emerging biblical-animist schema. In March 1870, at age thirty-one, he writes,

> I am sitting here in a little shanty made of sugar pine needles this Sabbath evening. I have not been at church a single time since leaving home. Yet this glorious valley might well be called a church, for every lover of the great Creator who comes within the broad overwhelming influences of this place fails not to worship as he never did before. The glory of the Lord is written upon all his works; it is written plainly upon all the fields of every clime, and upon every sky, but here in this place of surpassing glory the Lord has written in capitals. I hope that one day you will see and read with your own eyes.[19]

Here we see Muir's refined, but revolutionary, facility for recovering Christian piety's roots in the stirring marvels and rhythms of the natural world. At first glance, it may appear that Muir has taken leave of biblical faith. In this letter, he writes that he has not been to church even once since leaving his family's Wisconsin farm house ten years earlier. At this juncture, it would be easy to misunderstand Muir as losing his religion. But Muir is not breaking free of Christian identity in this letter; rather, he is transposing this identity into a new, in my terms, Christianimist key. Now Muir expands the definition of church to include the wild highland stretch he consecrates as his spiritual home in Yosemite Valley near Half Dome and Yosemite Fall. It is this "glorious valley" that is his subject of "worship"—it is this charged site, or "thin place" as we discussed earlier, where God is viscerally felt and worshipped therein. And it is not only this worshipful site but now *all* of creation, he writes, that is saturated with divine magnificence: the whole world is Muir's church just insofar as the "glory of the Lord is written upon all his works." The grand sweep of the entire natural order is a capacious and welcoming house of worship—a living site of reverence and awe continuously testifying to God's presence in all things.

It is important to understand here the important trope of *nature as sacred text* in Muir's thought. One day after his letter to David Muir extolling the ecclesial character of the Yosemite valley, Muir writes similarly, "God's glory is over all His works, written upon every field and sky, but here it is in larger letters—magnificent capitals."[20] Muir's mountain ecclesiology uses a biblical idiom to identify nature as the "book" or "text" on which is engraved God's astounding beauty. The church of nature is a book of towering pines, howling windstorms, and crystalline snow drifts wherein divinity is everywhere inscribed. In the Gospels, Jesus goes to the temple as a boy to learn from and teach scripture to the elders (Luke 2:41–52). Likewise, Muir, as a young man, enters into the temple of divine nature in order to see God and preach to the uninitiated the goodness of creation. The Psalmist says that "the heavens tell the glory of God, and the sky above proclaims his handiwork" (Psalms 19:1); Paul writes that "ever since the creation of the world, God's eternal power and divine nature . . . have been understood and seen through the things which are made" (Romans 1:20); and again, as we saw in Chapter 1, Paul avers that "the God who made the world and everything in it . . . is not far from each one of us, for 'In him we live and move and have our being'" (Acts 17:24, 27–28). Glossing the Bible's lofty animism, Muir quotes versions of these and similar texts and exults that divine glory is plainly written on all of God's works in big capital letters—and not just in one particular book or built place of worship.

In this regard, Muir subscribes to the scriptural metaphor of the *two books of revelation*, as this classical trope has come to be known.[21] From Tertullian and Origen to Galileo and Muir, many Western thinkers regarded the Bible and nature as revelatory texts in which each is best read and understood through the optics of the other. As we saw in Chapter 3, Hildegard valorized both "books" in this manner: her herbalist theology of *viriditas* used the Bible as a palimpsest in which traces of nature's medicinals could be seen in scriptural stories and images. Similarly, in 1866, at age twenty-eight, Muir writes, "I will confess that I took more intense delight from reading the power and goodness of God from 'the things which are made' than from the Bible. The two books harmonize beautifully, and contain enough of divine truth for the study of all eternity."[22] Referencing Romans 1:20, Muir enjoys more "delight" from "the things which are made" than from the scriptural canon, but he also claims that "the two books harmonize beautifully." In my judgment, Muir's point that the two books are complementary notes in the same piece of music is the critical point. Notwithstanding his comment that he prefers one book over the other, Muir's two-books theology is, arguably, the dominant framing device he uses to structure his understanding of the aboriginal unity between the natural world and the biblical texts throughout the course of his life. This framing device provides the binary tension that drives his expansive theological vision—the dialectic between earth and heaven, animality and divinity, and corporeality and scripturality. It is crucial to underscore that *because* Muir intimately knows and keenly values both books—sacred nature and holy writ—and sees no genuine conflict between their worldviews, he *therefore* is able to layer together both modalities in a point-counterpoint theological vision. The material world of God and the textual world of the Bible are a unified totality in Muir's thought. The book of nature and the Christian holy book operate as dual foci within a single ellipse, both of which lead to a deepening relationship between God and nature that is mutually renewing and reciprocally transforming.

Muir's contrapuntal theology of the two books is at play in a subsequent letter he writes to his brother, David, one month after his letter on the worldwide church of nature. In April 1870, he recasts the question of baptism in light of the power of sunlight, plant life, and running water in the church of the Yosemite landscape: "I was baptized three times this morning. 1st (according to the old ways of dividing the sermon), in balmy sunshine that penetrated to my very soul, warming all the faculties of spirit, as well as the joints and marrow of the body; 2d, in the mysterious rays of beauty that emanate from plant corollas; and 3d, in the spray of the lower Yosemite Falls. My 1st baptism was by immersion, the 2d by pouring, and the 3d by sprinkling. Consequently all Baptists are my brethren, and all will allow that I've 'got religion.'"[23] Alongside the sacrament of the Eucharist, baptism is arguably the central defining ritual of historical Christianity. Jesus insisted on his own baptism by John the Baptist (Matthew 3:13–17); the Ethiopian eunuch, one of Christianity's earliest converts outside the traditional holy lands, completed his transformation through baptism (Acts 8:26–40); and Paul says, in summary, that baptism is salvation, what he calls "dying with Christ" (Romans 6:1–11).

In biblical and extrabiblical Greek, the word *baptism* (*baptize*) means "to immerse," "to repeatedly dip," "to submerge," "to wash," "to cleanse," and "to practice ceremonial ablution." Muir is theologically astute about the three different methods of baptism practiced by traditional Christian denominations and uses this sacramental rite as a paradigm for giving meaning to this particular highland sojourn in April 1870. In his baptism letter, Muir correlates each of his sacred purification experiences on one particularly fine Yosemite morning—bathing in sunlight, soaking in the sight of various flower petals' circular formations, and being refreshed in the spray of a waterfall—with the different modes of baptism practiced by historical Christian churches: *immersion* (the baptismal candidate is completely submerged in water), *affusion* (the candidate has water poured over his or her head), and *aspersion* (the candidate is sprinkled with water). Structuring his Yosemite morning according to the classical Christian baptismal practices, Muir undergoes full "immersion" by bathing in sunlight, experiences "affusion" by absorbing the beautiful colors of native plants and flowers, and delights in the "aspersion" of cascading water in the lower Yosemite Falls.

For Muir, then, it is no small thing to describe his total submersion in nature's aesthetic bounty in salvific, baptismal language. A year after this letter to his brother, he writes similarly about baptism to Jeanne Carr, in April 1871, in the midst of a soul-refreshing trip to Yosemite Falls: "Oh, Mrs. Carr, that you could be here to mingle in this night-moon glory! I am in the upper Yosemite Falls and can hardly calm to write, but from my first baptism hours ago, you have been so present that I must try to fix you a written thought."[24] While Muir's phrasing, in this correspondence and in the earlier baptismal letter to his brother, is lighthearted and seemingly offhanded and his tender affections toward Carr are touchingly noteworthy, the broader theological intent of these dispatches should not be missed. Worship in nature's church—one of the two books of revelation—entails baptism into the wilderness of warm sunshine, floral bouquets, and cascades of water. Salvation is here and now and always-already present in the sheer gratuity of creation. *For Muir, then, nature is baptism. Nature is church. Nature is redemption. Nature is God.* The Gospel message is not extrasensory and otherworldly but fully sensual and erotically charged through its deep and abiding union with the living Earth. As early as 1871, Muir's *ad fontes* return to Christian faith's beginnings in the two books of creation and scripture is complete: Nature is God in all of God's gloriously sensual, pictorial, and liquid modes of renewal and salvation.

My point here is that the basic cast of Muir's thought is discernible at an early stage in his life in the manner he used scripture to shape and explain his journeys deep into the Yosemite wilds. In the same time period when he was writing David Muir and Jeanne Carr, he writes Carr once more, in February or March 1871, this time quoting Mark 1:12 ("And immediately the Sprit led [Jesus] into the wilderness"), and says, "'The Spirit' has again led me into the wilderness, in opposition to all counter attractions, and I am once more in the glory of Yosemite."[25] As Jesus is led in the Synoptic Gospels into a remote and uninhabited place in preparation for his public ministry, so also Muir sees himself as prompted by the Spirit to go into the wilderness and begin his ministry as a missionary for wild nature.

One contemporary translation of Mark 1:12–13, by Eugene Peterson, reads, "At once, this same Spirit pushed Jesus out into the wild. For forty wilderness days and nights he was tested by Satan. Wild animals were his companions, and angels took care of him."[26] Also led by the Spirit into the Yosemite hinterlands, Muir muses about the wild animals who were his constant companions therein—and doing so in a manner akin to Jesus' co-occupation with the untamed creatures of his own time. As was the case with Jesus, the backcountry for Muir is not to be shunned but, instead, enjoyed as a special site of divine habitation. In particular, in Muir's case, wild and remote Yosemite is the site where one of his favorites of all of God's beloved animal-children—the California grizzly bear—is free to pursue its own savage and joyous ends, always nurtured and cared for by the creator. Wild animals were Jesus' companions in the Palestinian desert; now the wild bears have become Muir's companions in the Sierra Nevada. On or around October 8, 1871, a few months after being Spirit-led into wilderness, Muir makes the following note in his journal:

> *Thoughts about Finding a Dead Yosemite Bear.* Toiling in the treadmills of life we hide from the lessons of Nature. We gaze morbidly through civilized fog upon our beautiful world clad with seamless beauty, and see ferocious beats and wastes and deserts. But savage deserts and beasts and storms are expressions of God's power inseparably companioned by love. . . .
>
> We deprecate bears. But grandly they blend with their native mountains. . . . Magnificent bears of the Sierra are worthy of their magnificent homes. They are . . . children of God, and His charity is broad enough for bears. They are objects of His tender keeping. . . . Bears are made of the same dust as we, and breathe the same winds

> and drink of the same waters. . . . And whether [the bear] at last goes to our stingy heaven or no, he has terrestrial immortality. His life not long, not short, knows no beginning, no ending. To him life unstinted, unplanned, is above the accidents of time, and his years, markless and boundless, equal Eternity.
>
> God bless Yosemite bears![27]

Here Muir writes about California bears in the philosophical vein of "deep ecology" but with a spiritual valence. Deep ecology philosophy postulates that all beings are goods unto themselves who have the right to live and flourish according to their own ends. The core presupposition of deep ecology—every life-form is an equal bearer of intrinsic value vis-à-vis all other life-forms and entitled to grow and blossom in its own way[28]—is the assumed premise of Muir's paean to the *Ursus horribilis* in the High Sierra. In this manner, Muir's ursine religion, as it were, inverts the spiritual value system that we use to elevate human beings as God's unique offspring and depreciate other beings as lesser than ourselves. We denigrate bears and other large mammals as "savage beasts," destined for quarantine and extirpation, but Muir lauds Yosemite's bears as "the children of God" who are "expressions of God's power" and "love" and "made of the same dust as we" are. In an animist vein, humans and bears are made of the same elements, sharing personhood and common identities as God's offspring. Anticipating the spiritual force of the latter-day philosophy of deep ecology, Muir alternately recalls Genesis 2:7, in which "God made man from the dust of the ground," but now writes that humans and bears alike are created out of earth; and he references Matthew 5:9, wherein Jesus says, "Blessed are the peacemakers, for they will be called children of God," but now Muir says the bears are God's blessed children and the fitting focus of God's tender keeping.

Muir's "blessing of the bears" journal entry makes perfect sense from a scriptural perspective and perhaps reflects his knowledge of the fantastic bestiary within the book of Job. Scholars generally agree that Job is the oldest book in the Bible (composed around the sixth century BCE) and reflects a theological sensibility that predates the priestly-scribal order of ancient Israel responsible for most of the biblical writings that followed after Job. It explores questions that have bedeviled philosophers and poets since time immemorial—questions such as the nature of God, the mystery of evil, and the place of humankind in the universe. In this register, Job speaks directly to one such perennial question, namely, Does creation consist of an ascending order of value among different natural beings, and, if so, which life-form carries the most value in this hierarchical order? Is it humankind or some otherkind? In biblical terms, are human beings the creator's honored favorites, or does this accolade belong to some other being? In a move that is counterintuitive to the Bible's presumed anthropocentrism, Job interrogates this problem by assigning high praise *not* to the human animal but to *another* animal—a powerful and massive land creature called "behemoth"—as the Almighty's primordial creation, namely, the "first of the great acts of God": "Look at Behemoth, which I made just as I made you; it eats grass like an ox. Its strength is in its loins, and its power in the muscles of its belly. . . . *It is the first of the great acts of God*—only its Maker can approach it with the sword. For the mountains yield food for it where all the wild animals play" (Job 40: 15, 19; emphasis added). In all of God's works, behemoth is first in creation. Job's claim is quite extraordinary in relation to what for many people is an intuitive sense that the human being is God's privileged and highest created life-form. Amazingly, Job says otherwise. Nevertheless, the book leaves unresolved the question of what precisely this behemoth (*behemeth* in Hebrew) is in the ancient cosmology. The scriptural depiction of this creature is of an enormous wild animal with strong core muscles that lives in the mountains and eats grass. Biblical commentators speculate that it could be a rhinoceros, hippo, elephant, or buffalo—or perhaps some other terrestrial mythological creature. But could it be a bear? The biblical account is too general to be sure, but pondering whether behemoth might be another name for the bear could make sense—and especially in the light of the Bible's frequent stories about bears' ferocious attacks (1 Samuel 17:34–37; 2 Kings 2:24) and their function as symbols of then-global political empires (Daniel 7:5). The biblical accounts of the great and terrifying bears of ancient Palestine and Muir's descriptions of the formidable grizzlies in the California high country fit together with a certain fearsome symmetry.

In the face of the traditional hierarchy of nature, in which the human being is privileged and made first in the order of things, Job counters that it is the colossal behemoth—meaning, perhaps, the bear—that is God's privileged and premier act of creation. In Job, and Muir, humankind is humbly bumped down the long and tangled Great Chain of Being that has held fast the metaphysical anthropocentrism of Western societies for centuries. In its stead, Job's behemoth, perhaps a predecessor of the wild bear later witnessed by Muir in the mountains, is now assigned an ancient, and contemporary, pride of place. "Magnificent bears are children of God" and "God bless Yosemite bears!" are the first and last lines of Muir's exclamatory hymn of praise to the Christianimist order of creation prefigured in Job's original cosmogonic vision.

Muir's exuberant and wide-ranging biblical citations continue apace throughout his life as a preacher of divinized nature. Remembering Jesus' words in the Sermon on the Mount, in which Jesus teaches his disciples to pray, asking, "Give us this day our daily bread" (Matthew 6:11), Muir writes, "Every purely natural object is a conductor of divinity, and we have but to expose ourselves . . . to be fed and nourished by them. Only in this way can we procure our daily spirit bread."[29] Harking back to Jesus' words of comfort and rest, "Come to me all of you that are weary and heavy laden, and I will give you rest" (Matthew 11:28), Muir entreats California industrial and farm laborers to come into the mountains for rest and renewal: "Come all who need rest and light, bending and breaking with over work, leave your profits and losses and metallic dividends and come."[30] And recalling that "Jesus went into the temple of God, and cast out . . . and overthrew the tables of the moneychangers" (Matthew 21:12), Muir rails against the wealthy political schemers who seek to exploit Hetch Hetchy as a commercial water source, writing bitterly, "Thus long ago a few enterprising merchants utilized the Jerusalem temple as a place of business instead of a place of prayer, changing money. . . . Ever since the establishment of the Yosemite National Park, strife has been going on around its borders and I suppose this will go on as part of the universal battle between right and wrong, however much its boundaries may be shorn, or its wild beauty destroyed."[31]

Toward the end of Muir's life, he deploys a particularly poignant biblical image to make his case for saving wild places. The Yosemite champion compares the cutting down of ancient trees in California's Sequoia redwood belt to Jesus' cathartic and sacrificial death in the New Testament. In what is today Calaveras Big Trees State Park, there still survive copses of dark reddish-brown Sequoia trees, some dating back to the first century CE and measuring upward to thirty feet in height. It is hard to imagine that these so-called millennial trees hail from the same time period as Jesus himself. In Muir's time, because the vulnerable Calaveras groves were owned by a private family and destined for the sawmill, he marshals the full resources of his dynamic scriptural rhetoric to stop the decimation of what he calls these sacred, noble king trees of the Sierra mountains.

Once again, the sweeping arc of Muir's biblical-animist project comes into full view. Like his pointed, lyrical letters and writings of the 1860s and 1870s as a young man exploring Yosemite for the first time, Muir, somewhere around the time of his death in 1914, pens "Save the Redwoods," a biting, sarcastic homily against the brainless stupidity of lumber barons wiping out an ancient forest for short-term financial gain. Muir notes that two of the great Calaveras trees have already been cut—and he cries out that more, in all of their "god-like majesty," are likely to follow. Recollecting Jesus' generous words of compassion in the midst of his own crucifixion, "Father, forgive them; for they know not what they do" (Luke 23:34), Muir imagines one of the two great Calaveras trees that had previously been felled reaching out, likewise, in love and clemency to its human torturers, even as Jesus had done so to his executioners: "This [earlier Calaveras] grand tree is of course dead, a ghastly disfigured ruin, but it still stands erect and holds forth its majestic arms as if alive and saying, 'Forgive them; they know not what they do.' Now some millmen want to cut all the Calaveras trees into lumber and money."[32] Muir deftly uses the trope of Jesus' crucifixion to stop the lumber-extraction madness in the High Sierras. As Golgotha sears the consciences of everyone who is complicit in the suffering of innocent victims, so Calaveras judges the idiocy of mercenary loggers who destroy God's temple for cash. Like the crucified one's pierced hands of mercy that extend to embrace all manner of malevolence, so also do the great scarred arms of the ancient Calaveras tree-person, in the wide compass of Christianimism, spread outward, not in judgment but in love, to enfold even the most unscrupulous mill bosses in mercy, sorrow, and absolution.

## *Sequoia Religion*

For some scholars of Muir, making sense of his biblically inspired nature writings is a difficult task. In general, Muir scholars do not question the intensity and frequency of his biblical exegesis in articulating his vision of sacred nature. No one is surprised, for example, that at age twenty-nine, in one of his earliest long treks, later narrated in his *A Thousand-Mile Walk to the Gulf*, Muir carried with him "the New Testament, Milton's *Paradise Lost*, a collection of Robert Burns' poems, and a journal."[33] It is difficult to dispute that under the surface of Muir's poetic imagination there runs a deep current of biblical language and imagery. But in the widening field of Muir studies, the question remains as to the significance and purpose of the Bible in shaping Muir's rhetorical ambitions.

The historians Frederick Turner and Evan Berry, for example, constitute one group of critics who regard Muir as a broad-minded eco-Christian who urges his readers to eschew otherworldly religious dogma and plumb instead the depths of divinity manifested everywhere within the natural world. In their mind, Muir saw no contradiction between biblical Christianity and environmental advocacy. For example, Turner writes, "Muir was first and foremost a Christian, and even if the fit of the faith was uncomfortable in places and had to be considerably altered to fit his own spiritual needs, it served well enough over time. Christianity might have its blindnesses, and he would define these sharply in coming years, but it was surely better than unbelief."[34] Analogously, Berry emphasizes the Christian resonances and biblical literacy characteristic of Muir's nature religion. He writes, "Syncopating ecological interconnectedness with Christian ideas about salvation and human purpose was already a mature topic of consideration in Muir's writings," and "Muir's scriptural style was of course an effective way to communicate the profundity of his message."[35] I agree with Turner and Berry: Muir is notable for marshaling the full spectrum of biblical nature imagery to articulate the joy of dwelling within, indeed worshipping, divinized nature.

But another group of Muir interpreters claim that the intrepid Scotsman renounced the Christian faith of his father and his youth and thereafter deployed biblical references only as a rhetorical stratagem aimed at satisfying his readers' religious sympathies. They aver that nothing in Muir's corpus points to a sincere desire to integrate biblical imagery and nature observation. When Muir writes, "Now all of the individual 'things' or 'beings' into which the world is wrought are sparks of the Divine Soul variously clothed upon with flesh, leaves, or that harder tissue called rock, water, etc."[36] or again when he imagines reassuringly the setting sun over Yosemite saying to "every stream and mountain, 'My peace I give unto you,'"[37] quoting exactly Jesus' words in John 14:27 to his disciples, these scholars contend that such sentiments are not a genuine expression of Muir's biblical naturalism, the emergence of a Christianimist hybrid in his thinking, to use my terminology, but a calculated plea to his readers to adopt his worldview by appealing to their religious beliefs and piety.

Three contemporary Muir scholars—Stephen R. Fox, Michael P. Cohen, and Bron Taylor—offer this latter take on Muir's project. Fox and Cohen start with the useful fiction that on or around age twenty-nine Muir initiated a lifelong split from Christian faith and the Bible. After that time, they claim, Muir utilized traditional devotional vocabulary as a tactical ploy, not as an exercise in religious conviction. Fox posits that "during this fall of 1867, in the course of working out his own philosophy Muir made a permanent break from Christianity. . . . He still read his New Testament—but more for the stately cadences of the prose than for theological ideas."[38] Fox's subsequent declaration concerning "Muir's apostasy from the faith of Jesus"[39] is matched by Cohen's similar biographical judgment: "since 1866 [Muir] had completely left his dependence on the Book of Books, and would never again accept the Bible as final authority."[40] Cohen continues that Muir had embarked on a sort of Taoist life path as a young man and had decided that from then on he would "reject the false and abstract doctrines of Christianity and learn his philosophy directly from Nature."[41] And what about Muir's seemingly frequent use of the Bible? According to Cohen, it was a sop he threw to the religious attitudes of his readers—a stylistic subterfuge for convincing people of faith to accept his nature philosophy: "Muir found it necessary to argue in terms that Christian people might understand. His text was Nature, but he could use a Biblical text if it would strengthen his argument."[42]

In spite of the surfeit of biblical references to God, Spirit, and the divine in Muir's work, Bron Taylor likewise agrees with Fox and Cohen that Muir's spiritual vocabulary was primarily strategic and political in nature: "[Muir] also regularly invoked God in his writings, including expressions of gratitude to the Deity for nature's beauty, which some readers have interpreted as evidence of theism. A more likely interpretation is that Muir thought it *politically useful* to use language that would be compelling to the various publics he sought to enlist in the cause of environmental protection, which included romantics, transcendentalists, and theists."[43] Of all the misleading readings of Muir, I find Taylor's interpretation to be particularly specious. In a work on "green religion" and "nature spirituality"—the key terms in the title and subtitle of his book, respectively—Taylor dismisses the groundbreaking ecotheology of (arguably) America's most seminal environmental activist and author as a function of "politically useful" language. My instinct is that lying behind Fox's, Cohen's, and Taylor's strained interpretations of Muir's biblical rhetoric as an exercise in political expediency is their unexamined premise that sacred nature spirituality and biblical religion are somehow antithetical to each other. Taylor writes that "Muir's tendencies were first and foremost animistic,"[44] as if such a claim disqualifies Muir's earnest desire, or so it seems to me, to root his animism in the rich soil of the biblical imaginary.

In my judgment, the scriptural rootedness of Muir's passion for sacred nature is not incidental to but, rather, dynamically generative of this passion in the first place. Taylor et al. assume that Muir is a *post*-Christian animist, ruling out any return on Muir's part to the biblical heritage of his youth, whereas what seems more plausible is that Muir sees no inherent contradiction between his nature spirituality and his lifelong biblical fidelities. As I have sought to show here, Muir's eco-mysticism is both enhanced and deepened by his delight in and ease with biblical and Christian ideas such as God, baptism, Eucharist, salvation, goodness, sacrifice, and forgiveness. Simply put, Muir is both an *animist* and a *biblicist*—he sees no conflict, to employ the dual grammar of his worldview, between belief in Spirit within nature, on the one hand, and joy in the presence of God in creation, on the other.

In this vein, there is no better example of Muir's lively alchemy between deified nature and biblical theology than his now-famous "Lord Sequoia" letter to Jeanne Carr, written on or around 1870. This strange, erotic, effusive "woody gospel letter," as Muir calls it, gives the time of writing as "Nut Time" in the date line and notes the place of writing as "Squirrelville, Sequoia Co." in the address line. The letter is chock-full of odd spelling, grammatical license, unique neologisms, and an excessive and exuberant style distinctive even for Muir's signature epistolary aesthetic. Also unique: Muir makes a quill from a Yosemite golden eagle feather and the sap of a giant Sequoia tree (perhaps in combination with a solvent or stabilizer) as the writing stylus and ink, respectively, with which to compose the letter.[45] Not only the rhetorical modalities but also the physical means by which Muir writes this mountain missive are "Sequoical" and "terrestrialized," to use his newly coined expressions in this letter. As the best example of Muir's biblical-animist amalgam, I can only do justice to the full power and magic of this feather-pen, purple-Sequoia-ink letter by quoting it in full.

*To Mrs. Ezra S. Carr*
SQUIRRELVILLE, SEQUOIA CO.
*Nut Time*

DEAR MRS. CARR:

Do behold the King in his glory, King Sequoia! Behold! Behold! Seems all I can say. Some time ago I left all for Sequoia and have been and am at his feet, fasting and praying for light, for is he not the greatest light in the woods, in the world? Where are such columns of sunshine, tangible, accessible, terrestrialized? Well may I fast, not from bread, but from business, book-making, duty-going, and other trifles, and great is my reward already for the manly, treely sacrifice. What giant truths since coming to Gigantea, what magnificent clusters of Sequoic *becauses.* From here I cannot recite you one, for you are down a thousand fathoms deep in dark political quagg, not a burr-length less. But I'm in the woods, woods, woods, and they are in me-ee-ee. The King Tree and I have sworn eternal love—sworn it without swearing, and I've taken the sacrament with Douglas squirrel, drunk Sequoia wine, Sequoia blood, and with its rosy purple drops I am writing this woody gospel letter.

I never before knew the virtue of Sequoia juice. Seen with sunbeams in it, its color is the most royal of all royal purples. No wonder the Indians instinctively drink it for they know not what. I wish I were so drunk and Sequoical that I could preach the green brown woods to all the juiceless world, descending from this divine wilderness like a John the Baptist, eating Douglas squirrels and wild honey or wild anything, crying, Repent, for the Kingdom of Sequoia is at hand!

There is balm in these leafy Gileads—pungent burrs and living King-juice for all defrauded civilization; for sick grangers and politicians; no need of Salt rivers. Sick or successful, come suck Sequoia and be saved.

Douglas squirrel is so pervaded with rosin and burr juice his flesh can scarce be eaten even by mountaineers. No wonder he is so charged with magnetism! One of the little lions ran across my feet the other day as I lay resting under a fir, and the effect was a thrill like a battery shock. I would eat him no matter how rosiny for the lightning he holds. I wish I could eat wilder things. Think of the grouse with balsam-scented crop stored with spruce buds, the wild sheep full of glacier meadow grass and daises azure, and the bear burly and brown as Sequoia, eating pine-burrs and wasps' stings and all; then think of the soft lightningless poultice-like pap reeking upon town tables. No wonder cheeks and legs become flabby and fungoid! I wish I were wilder, and so, bless Sequoia, I will be. There is at least a punky spark in my heart and it may blaze in this autumn gold, fanned by the King. Some of my grandfathers must have been born on a muirland for there is heather in me, and tinctures of bog juices, that send me to Cassiope, and oozing through all my veins impel me unhaltingly through endless glacier meadows, seemingly the deeper and danker the better.

See Sequoia aspiring in the upper skies, every summit modeled in fine cycloidal curves as if pressed into unseen moulds, every bole warm in the mellow amber sun. How truly goodful in mien! I was talking the other day with duchess and was struck with the grand bow with which she bade me good-bye and thanked me for the glaciers I gave her, but this forenoon King Sequoia bowed to me down in the grove as I stood gazing, and the highbred gestures of the lady seemed rude by contrast.

There goes Squirrel Douglas, the master-spirit of the tree-top. It has just occurred to me how his belly is buffy brown and his back silver gray. Ever since the first Adam of his race saw trees and burrs, his belly has been rubbing upon buff bark, and his back has been combed with silver needles. Would that some of you, wise—terrible wise—social scientists, might discover some method of living as true to nature as the buff people of the woods, running as free as the winds and waters among the burrs and filbert thickets of these leafy, mothery woods.

The sun is set and the star candles are being lighted to show me and Douglas squirrel to bed. Therefore, my Carr, good-night. You say, "When are you coming down?" Ask the Lord—Lord Sequoia.[46]

The whole letter is a theo-poetic feast—a running biblical-cum-naturalist commentary with distinctive shifts in extravagant terminology pregnant with significance. Here let me highlight the biblical source material Muir borrows from in order to excavate the multiple layers of religious and ecological meaning buried in the letter. The first line of the letter provides a clue to the sense of the letter itself—as well as to Muir's overall synthesis of scriptural and environmental themes in the full body of his work in toto. "Behold the King in his glory, King Sequoia!" is a reference both to Isaiah 33:17 ("Behold the King in his beauty!") and the constant accolade to "the King of glory" in the twenty-fourth Psalm. At first glance, it may seem that the referent of this royal paean has seemingly shifted from the immortal and invisible God of the Hebrew prophets to the Sequoia trees of the Yosemite mountains. But is this a shift at all?

If I am right that Muir's theology, in this letter and in general, posits all things as a microcosm of divinity—if, as he says, "all of these varied forms [of life], high and low, are simply portions of God radiated from Him as a sun, and made terrestrial by the clothes they wear"[47]—then, alternately and at the same time, without division and without confusion, praising the biblical ruler of the universe, on the one hand, and King Sequoia in his glory, on the other, consists of one and the same adulatory gesture. In Muir, there are not *two* orders of being—the supernatural and the natural—but only *this* world in all of its Spirit-saturated glory and beauty.[48] In this regard, note that the letter opens with an invocation of "King Sequoia" and finishes with a benediction to "Lord Sequoia." Following, therefore, the arc of this arboreal-monarchical language from the letter's beginning to its end renders Muir's equation between the biblical

God of creation and the forestal God of Yosemite full and complete. With this interpretive key in hand, the meaning of the rest of Muir's "woody gospel letter" opens up.

After successive tributes to King Sequoia, Muir makes a series of lyrical claims that transpose foundational biblical terms—Eucharist, oneness, sacrifice, repentance, and salvation—into a new environmental key that realizes the true ecological intent of the original scriptural terms. This move begins with Muir's ecstatic declaration of his total physical and mystical union with the forest itself: "I'm in the woods, woods, woods, and they are in me-ee-ee." Here the reader can feel Muir's semantic range bursting at the seams; he extends the alliterative reach of his voice to its breaking point in order to realize his full coinherence with the woods. This declaration is quickly followed by another: "The King tree and I have sworn eternal love—sworn it without swearing, and I've taken the sacrament with Douglas squirrel, drunk Sequoia wine, Sequoia blood, and with its rosy purple drops I am writing this woody gospel letter." In this instance, Muir borrows the biblical language of God swearing God's eternal love for the chosen people to illustrate the sacred forest covenant that Muir now enters into with King Sequoia. As a living member of this eternal bond, Muir recasts one of the two central sacraments of Christianity, the celebration of the Lord's Supper, in terms of eating the flesh of a Douglas squirrel and drinking the blood of a Sequoia tree. (Whether by having "taken the sacrament with Douglas squirrel" Muir means that he is literally eating squirrel flesh is unclear—though the following clause about having "drunk Sequoia wine" and subsequent vision of "eating Douglas squirrels" make this carnivorous option more likely.) In any event, the forest ritual that Muir describes points to a familiar Christian parallelism: by ritually consuming the prescribed elements of flesh and blood—in this case, squirrel meat and wood sap—Muir is eucharistically eating the fleshy body and drinking the sappy blood of incarnate divinity.

The next paragraph is initially marred, however, by Muir's dismissive reference to Indians: they "instinctively drink" Sequoia juice "for they know not what." Earlier we read Muir's comment that "perhaps if I knew [the Indians] better I should like them better." It appears, however, that Muir did not consistently expend the effort to learn about first peoples' sophisticated local epistemologies concerning the very Sequoia flora and fauna he rhapsodizes over in this paragraph. At times, he does praise the Ahwahneeche and other Native people for being the first to articulate the animist worldview central to his vision: "To the Indian mind all nature was instinct with deity. A spirit was embodied in every mountain, stream, and waterfall."[49] At other times, however, he devolves into familiar assumptions about the deficiencies of Indigenous apperception regarding Yosemite's many natural gifts.

In spite of this lapse, Muir continues in this paragraph of the letter with a powerful and familiar shape-shifting motif in Muir's own self-understanding: John Muir is a new John the Baptist. By preaching with abandon the salvific power of full immersion in the Yosemite backcountry, Muir wears the mantle of his namesake and continues the ancient John's wilderness mission of repentance, baptism, and redemption. Muir's self-referential allusions to this eponymous green prophet are an idée fixe in his writing, as in this journal entry, to give just one example, from October 1871: "Heaven knows that John the Baptist was not more eager to get all his fellow sinners into the Jordan than I to baptize all of mine in the beauty of God's mountains."[50] Muir's intertextual references to the Gospels' stories of John the Baptist (Matthew 3, Mark 1:2–8, Luke 3:1–18, John 1:19–28) are impossible to miss in this quote and his woody gospel letter. As the one John descended from the wilderness of Judea preaching that the kingdom of God is at hand, so the other John descends from the wilderness of Yosemite preaching that the kingdom of Sequoia is at hand. As the one John wore camel's hair and ate wild locusts and honey, so the other John wears cotton and wool hiking suits and eats Douglas squirrels and wild honey. And as the one John said repent to all of the sinful denizens of Judea and baptized them in the cleansing Jordan River, so the other John says repent to all of the so-called civilized inhabitants of "the juiceless world" and seeks to fully baptize them in the beauty of the High Sierras. Driven by the Holy Spirit into the California wilds, Muir emulates his biblical forbearer, who was known as a "voice crying out in the wilderness" (Matthew 3:3), and like the age-old John the Baptist, the new John the Baptist also cries out "repent!" so that his hearers might experience healing and salvation through his woody gospel letter and similar writings.

Finally, quoting Jeremiah 8:22's appeal to a "balm in Gilead"—a famous tree resin, from an area in contemporary Jordan, used as a salve in traditional Hebrew medicine[51]—Muir claims a restorative "balm" for "all defrauded civilization; for sick grangers [i.e., farmers' organization] and politicians" in the primeval Yosemite forest, in the "leafy Gileads" of the Sequoia highlands. This healing relationship to the land is a sacramentally visceral, erotic, and inebriating experience; it is a way of being "drunk" with sacred Sequoia juice by hungrily slurping and sucking into one's mouth the sensuous beauty of the forest. With vertiginous delight, Muir declaims God's incarnation *ad litteram*: he orates a rapturous vision of a fleshy religion, an earthen-intoxicated-therapeutic-sexual religion, as it were. As well, Muir's woody queer sermon operates in a homophilic register. In John the Baptist–like hortative ecstasy, Muir preaches an eco-homoerotic altar call: "Come suck Lord Sequoia and be saved."[52] Effusively, Muir's green Christianity is not a dead belief system of desiccated bodies and dry bones but a flesh-loving spirituality of orgasmic joy in which the parched enjoy refreshment, the sick are healed, and the lifeless are made new. This is a religion of physical renewal, sexual healing, emotional integration, and spiritual vivification. Muir—or "John of the Mountains," as his early-twentieth-century chronicler Linnie Marsh Wolfe called him[53]—now sees himself as God's libidinal lure into the mountains: *Repent of your indifference and hostility to nature's bounty and come follow me into the loving embrace of forests and rocks, waterfalls and meadows, great cranes and water ouzels, and here in the green-brown woods, you will find rest for your bodies and peace for your souls.*

### *"Christianity and Mountainanity Are Streams from the Same Fountain"*

In *Confessions*, Augustine frequently refers to God as a loving fountain, glossing Psalm 36. God is an awe-inspiring reservoir of tender mercy, compassion, wisdom, and forgiveness. Amid life's turmoil, Augustine writes, he often forgets God and takes leave of God's offers of pardon and sympathy, but then he returns to God and slakes his thirst at God's fountain—the "well of life," as he calls it.[54] Augustine writes, "How much did I love you. I went astray, and remembered you. I heard your voice behind me bidding me return, and scarcely did I hear it for the tumults of the unquiet ones. And now, behold, I return burning and panting after your fountain. Let no one prohibit me; of this I will drink, and so have life."[55] Augustine burns and pants after this wondrous headwater—the fountain of mercies, faithful and abundant, that does not depend on any external source for its ever-flowing generosity and blessings. This fountain of life pours forth grace and well-being for all who drink from its eternal springs.

In a letter from June 1872 to Catharine Merrill, an Indianapolis university professor and Civil War historian, Muir uses the same Augustinian image—all good things come from God's perennial fountain—to explain the relationship between nature's sacred bounty and the biblical witness to Jesus' love for all.[56] It is possible, but unlikely, that Muir was well versed in Augustine's *Confessions*, but it is entirely possible that he was aware of Psalm 36 or similar scriptural texts that deploy divine fountain imagery. For Muir, he learns from Psalm 36 inter alia that both modalities—nature's bounty and Jesus' love—bubble forth from the same universal spring. The loving fountain that is God in Jesus, Muir says in the letter to Merrill, is the same fountain that wells up from the rocks and waters, mountains and trees, and fields and skies that encircle and ensoul all of life. Muir's generosity of spirit is evident in his reply to Merrill's criticism of his all-encompassing Christianimist vision. Like Muir's censorious father, who lambasted his son in 1874 for knowing full well that his nature writing "was not God's work," although he "seem[ed] to think [he was] doing God's service,"[57] Merrill scolds Muir for betraying Christianity's anthropocentric message by extending the horizon of God's saving love beyond the human community to the more-than-human world. Politely, Muir replies that divinity is never ending and unbounded and that he hopes that Merrill will join him for an extended sojourn in the mountain temple of Yosemite. In the mountains, he would like to change her mind so that she could return to university life with a new and expansive theological worldview after being saturated by God's illimitable love. Muir writes,

> I wish you could come here and rest a year in the simple unmingled Love fountains of God. You would then return to your scholars with fresh truth gathered and absorbed from pines and waters and deep singing winds, and you would find that they all sang of fountain Love just as did Jesus Christ and all of pure God manifest in whatever form. You say that good men are "nearer to the heart of God than are woods and fields, rocks and waters." Such distinctions and measurements seem strange to me. Rocks and waters, etc., are words of God and so are men. We all flow from one fountain Soul. All are expressions of one Love. God does not appear, and flow out, only from narrow chinks and round bored wells here and there in favored races and places, but He flows in grand undivided currents, shoreless and boundless over creeds and forms and all kinds of civilizations and peoples and beasts, saturating all and fountainizing all.[58]

Muir's hope is that Merrill will learn from the book of divine nature, that she will return to other "scholars with fresh truth gathered and absorbed from pines and waters and deep singing winds." This "fresh truth" entails the insight that nature's elements are "words of God" because all of us, so-called inanimate and animate beings alike, all "flow from one fountain

Soul." There are no sharp dividing lines in nature that separate God and this-worldly existence. There are no hard-and-fast distinctions that favor Christians and their beliefs as "nearer to the heart of God" than the non-Christians, and the nonhumans for that matter, as somehow distant from and isolated from God's loving heart.

For Muir, in summary, world-denying belief in the Christian God is the betrayal of genuine faith in the true God. He argues that the religion of so-called good men is unfaithful to the Gospels' portrayal of God's boundless love for all of us, good or bad, religious or nonreligious, human or animal or vegetal alike. Christianity, in its current "creeds and forms," is the betrayal of Christianity's core insight that "all are expressions of one Love." Thus, born and bred in the same river of divine love, "Christianity and mountainanity are streams from the same fountain," as Muir writes in another letter to another friend in January 1873, deploying again the same spring-water imagery he used with Catharine Merrill in 1872.[59]

But today the Christian religion has lost its way, and belied its own best insights, by taking flight from the mortal world of the flesh and shifting its gaze skyward to an immortal world of disembodied bliss. For Muir, Christianity, in a word, is still not Christianity. Instead, it has traded its birthright as the religion that proclaims that God became flesh and dwells among us for an otherworldly belief system that celebrates institutional power and hidebound creeds in place of joyous solidarity with the Earth and all of its blessed inhabitants. In my reading of Muir, this is the latter-day John the Baptist's continuous jeremiad against Western Christianity's antiworldly theology. In turn, it is the consistent message of this book: God's boundless compassion, especially in its many and varied ecological expressions, embraces all of us in the warm heart of nature's church, calling to us to worship therein and to protect this worldwide green temple as our loving home and common destiny.

CHAPTER 5

# On the Wings of a Dove

## *Sagebrush Requiem*

During my childhood in Southern California, I slept under a window that opened out to a series of foothills covered with sagebrush. There is much I remember about those foothills and the wider natural world I encountered there—the scamper of road runners, the howl of coyotes, the red flash of prickly pears amid the cactus. But what I remember most is the bittersweet fragrance of sagebrush wafting through my window at night, especially after a rare rainstorm.

Sagebrush is a woody, scrubby plant found throughout much of the western United States, among other similar Mediterranean climates. It grows wild in arid grasslands and desert regions. It is also cultivated for its ornamental flowers as well as its medicinal and herbal properties. I use sage in tea and recipes such as cornbread stuffing at Thanksgiving and paired with lemon in roasted chicken. It has long been considered one of the principal herbs for general cooking. Paul Simon and Art Garfunkel's 1966 ballad "Scarborough Fair" is a remake of an ages-old British melody that assigns special culinary status to sage as one of the four primary herbal spices used in English recipes. In "Scarborough Fair," sage is highlighted in a moving refrain that winds through the ballad:

Are you going to Scarborough Fair?
Parsley, sage, rosemary, and thyme;
Remember me to the one who lives there,
She once was a true love of mine.

Akin to sage's recurring role in the poignant chorus of the Simon and Garfunkel song, the sage of my youth was everywhere in the warm drafts and shifting winds of the California chaparral. Its peppery, pungent scent signaled the vitality of a robust ecosystem of Spanish oaks, horned toads and rattlesnakes, and wild manzanita, poppy, and lilac flowers. Not everyone likes the penetrating fragrance of sage—its strong smell can seem acrid and bitter to some—but I experienced sage as a heady mountain perfume that defined my boyhood foothills as a wild and untrammeled place. This was a place that had not yet been colonized by the shopping malls and freeway grids that characterized the rest of urbanized California. As a child, my senses were alive with a kind of scrubland aromatherapy through my daily immersion in an atmospheric bath of sweet-smelling sage. At night under my bedroom window, I enjoyed how the day's stress would drain from my body, and I would relax as I absorbed the scented breezes coursing through my room.

My father still lives in the house where I was raised—a house that he and his friends built in 1964—but when I return to my early home, I can no longer take in great drafts of sagebrush-scented air. In my youth, the air was aromatically textured depending on which way the winds blew and whether there had been rain or fog that day or the day before. But now, regardless of the passing weather, the air is odorless and thin; it does not carry the sagey, zesty smell of wild plants as it used to. While sage is still abundant in the West, its habitat is increasingly threatened by overdevelopment and the widespread use of herbicides. Sage plants are shallow rooted and easy to disturb. Once impacted, they struggle to reestablish themselves in soil regions that can no longer support fibrous root systems that lie close to the surface of the ground. I have heard that the sense of smell forms memories early in life that are the strongest recollections available to human beings. Many years later, my recall of the spicy scent of sage is still sharp and distinct. My memory is of a fragrant plant whose spiritual, medicinal, and gustatory properties are woven together in a sensual experience of deep pleasure and reassuring comfort and joy. And my sadness is that the daily delight I experienced drinking in the pungent aroma of this healing herb has been largely attenuated, if not soon wiped out altogether, by the expansive sprawl of the California Southland.

## *Is Earth a Living Being?*

My early immersion in sweet-smelling sage taught me that the Earth is an all-encompassing gift with the potential to heal and restore all of its inhabitants. My boyhood sage baths showed me that the world is an enchanted place where natural forces can provide daily pleasure and renewal. The Genesis creation story's hymn-like refrain, "And God saw everything that God had made, and behold, it was very good" (1:31), signals the buoyant grace and beauty within the natural world.[1] But this everyday splendor, in desert sage and across the planet, is under siege as human communities continue to lay waste to the goodness and bounty of creation.

Why are desert sage ecologies, and most other planet-wide ecosystems, in danger of collapse today? I suggested in Chapter 2 that human-caused climate change is the basis of both rising global temperatures and the massive depredation of species and habitats in the current setting.[2] Unsustainable, industrial civilization has dumped billions of metric tons of carbon into the atmosphere through fossil fuels burning. Heavy consumption of coal, oil, and natural gas, along with continued deforestation since the start of the industrial age, is causing global temperatures to escalate astronomically—anywhere from three to seven degrees Fahrenheit by 2050 to as high as seven to ten degrees by the century's end. While certain species of cockroaches and sea squids might survive this warming, such temperature rises are objectively apocalyptic for human well-being and untenable for most of planetary life as we know it. As Earth becomes hotter and drier, the relentless climb in global temperatures is degrading land masses and terrestrial life-forms as well as the chemical composition of the world's oceans. Low-lying islands such as the Maldives and areas such as South Florida are slipping underwater. The world's remaining glaciers are fast disappearing, with all of the arctic sea ice in the North Pole currently slated for oblivion. The extinction rate of amphibians, such as frogs and salamanders, is now forty-five thousand times higher than their normally expected rate of extinction. And our beautiful blue oceans, as erstwhile great carbon sinks that had been capturing excess carbon dioxide emissions in the atmosphere, are being rendered more acidic and inhospitable to diverse sea life, with as many as one-third of all coral reefs becoming bleached out or headed soon for annihilation. The good earth of the Genesis creation hymn has become an ecology of fear driven by planetary abuse. *Sic transit mundus.*

We have entered the fateful epoch of the "Sixth Great Extinction"—a geological time period similar to the last mass extinction event, when the dinosaurs were wiped out tens of millions of years ago. Vertiginously, we are climbing a dangerous staircase of global-warming-driven "tipping points"—catalytic chain-reaction events, such as melting permafrost, that could trigger widespread and sudden catastrophe within the heretofore self-regulating global climate system.[3] Teetering on the edge of these irreversible impacts on planetary ecosystems, we stand watch over the steady and inexorable demise of our plant and animal cousins and their landed surroundings—including, perhaps some day soon, the passing of the aromatic sage fields that populated the postwar Southern California of my youth.[4]

As the planet becomes hotter and cascading waves of species-level extinctions are the result, our reliance on fossil fuels continues apace. This carbon dependence stems from our shifting attitudinal postures toward the natural world. As I asked in Chapter 2, is Earth a fragile gift to be loved and protected, or is it a ready supply of energy and materials to be used and exploited? Earth, to use Martin Heidegger's formulation that we saw earlier, has become an extensive "standing reserve" of inexhaustible power for modern industrial development, and the root metaphors we employ to describe the natural world belie any hope we might have of extricating ourselves from our fundamentally abusive dispositions toward the life-giving systems on which we all depend. As standing reserve, Earth is not a "living being" or "feeling organism" with its own subjective moods and affective propensities. As raw material, Earth cannot feel pain or experience loss or undergo the suffering, some people claim, that only we humans and other conscious life-forms can feel. Our techno-supply vocabulary for Earth has effectively rendered our living planet numb and silent—a dead zone of inert matter, a fixed deposit of energy to fuel commercial development at all costs. As Robin Wall Kimmerer says, "In English, we speak of the land as 'natural resources' or 'ecosystem services,' as if the lives of other beings were our property."[5] As a lifeless thing, as an impersonal, mechanized repository of useful materials, Earth, in the terminology we use and with which we feel most comfortable, is now, in its most basic essentiality, a "resource" of "services" to supply the needs of human society—or, perhaps more accurately in a market-driven economy, a "commodity" to be bought and sold in the financial marketplace, like toothpaste or pork futures or stock options.

My aim throughout this project has been to counter this instrumental model of nature by *re-enchanting* Earth as an animate being, a living soul, a spiritual life force that feels a type of joy and suffers a certain kind of sorrow and loss in a manner analogous to what we also feel. In this attempt, my hope has been to empower our collective desire to heal Earth's suffering by rekindling a spiritual vision of our biotic and abiotic kinfolk as revered members of a unified, blessed family. While Christian theology has focused on human suffering in light of Jesus' passion, my goal has been to expand this horizon of concern to include the more-than-human others who also suffer and often do so at the hands of their human oppressors. My case is that once we forge an emotional bond between ourselves and nature—once we develop a heartfelt kinship between our kind and all other kinds as hurting and vulnerable members of a common family—then we will have the vision and energy to enter the public fray, to bind up the wounds of an injured planet, and to fight the long-term battle to save our own and other species as well.[6]

But to say that "Earth suffers," at first glance, will strike many of us as a contradiction in terms. How can the inanimate material world be said to undergo pain and suffering when, by definition, it is not a living, breathing organism as we are? Are not rocks, bodies of water, trees, and other plants rightly classified by biologists and philosophers as nonsentient matter because they lack conscious self-awareness and, as such, cannot experience pain and loss? The problem with such categorizations, however, is that they do not make sense either from an *ecosystemic* or a *neoanimist* perspective. Understood holistically, all members of the life web are self-organizing beings with their own moods and traits, power and agency, and corporeal vibrancy and evolutionary trajectories. In this vein, all things possess value all their own as vital contributors to diverse bionetworks and developmental complexity. It follows, therefore, that the everyday ontological binaries we use to organize our everyday sensible experience—binaries such as life/matter, organic/inorganic, living/nonliving, and sentient/nonsentient—are really only the self-justifying oppositions we deploy to rationalize hierarchically our own status in the scheme of things, not actual descriptions of the way things are. In effect, our typical dualisms function as classificatory stratagems that elevate us human beings and our ilk as self-realizing beings over and against all other lifeforms as unfeeling things.

I am suggesting that our commonplace taxonomies blind us to how the places, things, and elements around us are also living organisms with emotional registers all their own. Take, for example, the stone wall that runs along the level of my eye outside my study window as I write these words. Fixed and impassive, how could this squat, rocky wall be anything other than lifeless matter? In what sense could it be said to be a living, feeling being with dispositions and moods like the rest of us? My rock wall is made from Wissahickon schist, a beautiful and, at one time, ubiquitous local stone, flecked with quartz and mica, that has given Philadelphia and its surrounding architecture a uniformly earth-toned and stolid appearance. But while Wissahickon schist is aesthetically pleasing, in what sense can it be said to be alive? In response, let me suggest the following: the rocks in my wall are living things—as are the rocks strewn across the stony face of the planet—precisely because they are vital structural elements in the geochemical processes that support my and my family's existence in our Swarthmore home. My seemingly inert and immobile rock wall is actually part of a living, swirling ecosystem that energizes everything around it with interlocking vitality. Covered in lichen and microorganisms I cannot see, my stony exclosure holds together the teeming community of a/biotic life-forms that sustain my immediate niche within the larger eco-zone I inhabit. By controlling soil loss through sediment trapping, for example, it holds steady much of the biomass that ensures the well-being of my and my family's household. This biomass, including my yard's surrounding thicket of trees, shrubs, and groundcover, also plays a role in Earth's carbon cycle as one of the many links in the photosynthetic food chains that make all planetary life possible, in my bioregion and elsewhere. Among other critical functions, the absorption of carbon dioxide at my particular home site, and the corresponding production of oxygen, now stabilized by the wall outside my study window, is essential to my and my family's, and all other beings', survival.[7]

Are rocks, then, not dead things but vital members of the life web necessary for existence? The paleontologist James Lovelock argues for the intrinsic value of all of Earth's living elements—including, by implication, the Wissahickon schist in my rock wall—in the service of the functional integrity of the biosphere writ large. Lovelock theorizes that the planet is a "superorganism" in which all of its biological, physical, and chemical components are "alive" and necessary for the support and regulation of global biodiversity. Lovelock calls the living Earth "Gaia," named after the ancient Earth goddess of the Greeks, to signal the quasi-mystical powers of worldwide biochemical interactions among air, waters, trees, and rocks to create the ideal living conditions—including the ideal climate—for all inhabitants of the planet.[8] He calls this "the Gaia hypothesis" and frames it this way: "The entire range of living matter on Earth, from whales to viruses, from oaks to algae, could be regarded as constituting a single living entity, capable of manipulating Earth's atmosphere to suit its overall needs and endowed with faculties and powers far beyond those of its constituent parts."[9]

According to Lovelock, then, our particular human role in the biosphere is to *understand* how Earth's or Gaia's biophysical interactions create a steady state fit for life and then to *support* the capacities of this "single living entity" to maintain optimal ecosystem functionality for diverse communities of species. Lovelock writes, "The more we know, the better we shall understand . . . the consequences of abusing our present powers as a dominant species and recklessly plundering or exploiting [Earth's] most fruitful regions."[10] In reference to Lovelock, my corresponding point is that when we devolve into "abusing our present powers" and degrade the abilities of Gaia's interweaving elements to achieve their natural ends—in other words, when we cause any of the constituent members of diverse ecosystems to suffer harm—then we do injury to the vital organisms and processes that make our self-regulating planetary life system generative and sustainable. It is in this sense, therefore, that we can say that when we assail Gaia's ecosystemic balance, we are causing Earth, as an organic being, as a "single living entity," to quote Lovelock, to suffer harm, to feel pain, and to undergo trauma.[11]

## *Suffering Earth*

Is it possible to maintain that the wider, environing Earth we inhabit, as a living being, is able to feel grief and suffer injury? Understood animistically, I am suggesting not only that Earth systems science, understood holistically in the manner of Lovelock's organicism, is advancing just such a claim today but also that a searching interpretation of the Bible, now read from the perspective of Earth as a subject unto itself, will make the same argument. From a scriptural vantage point, the model of Earth as a purposeful, living organism with the capacity to feel and suffer is a green thread that runs through, and ties together, the entirety of the Hebrew Bible and New Testament. In the Bible, Earth is a vital "actant," to borrow a term the from social theorist Bruno Latour,[12] with its own affective tendencies and relational capacities. Like Lovelock's Gaia, Earth is a living, sentient, vibratory, purposeful, generative agent of its own action—not a dead thing—with its innermost moods, voices, feelings, and conscious states of being. Pulsating, germinating, evolving, producing—Earth embodies, in its deepest subjective personhood, a rich and textured emotional life full of sorrow and joy, mourning and hope, and groaning and laughter within the narrative cycles of the scriptures.

An early biblical example of Earth's personal agency and affective range surfaces in the opening pages of the book of Genesis, where God laments Earth's suffering in the light of Cain's murder of his brother Abel. In Genesis, Cain takes Abel to a nearby field in order to kill him, watches his lifeblood flow into the ground, and then purports not to know of the body's whereabouts, after which the Lord queries him:

> Then the Lord said to Cain, "Where is your brother Abel?" He said, "I do not know; am I my brother's keeper?" And the Lord said, "What have you done? Listen: Your brother's blood cries out to me from the ground. Now you are cursed from the ground which has opened its mouth to receive your brother's blood from your hand. When you till the ground, it will no longer yield to you its strength; you will be a fugitive and a wanderer on the earth." (Genesis 4:9–12)

Here Earth is not dumb matter, an inanimate object with no capacity for feeling and sentiment, but a spirited and vulnerable living being who experiences the terrible and catastrophic loss of Abel's death. Its heart broken and its mouth agape, Earth "swallows," in the text's startling imagery, mouthfuls of Abel's blood, "receiving into itself" the dark stain of the young herdsman who is the first victim in the Bible's long and ugly catalogue of familial horror stories. Bubbling up from the red earth, Abel's cries signal not only that Cain has murdered his brother but that he has done lasting, perhaps irreparable, violence to Earth as well. Aggrieved and bereaved, Earth weeps or shouts—the text says the blood-soaked ground "cries out" to God—at the terrible harm Cain has done to his only sibling and to the land as well. In murdering his brother, Cain inflicts pain on the nurturing earth ("Your brother's blood cries out to me from the ground"; Genesis 4:10) that gave to him and to his family their first life ("God formed the human from the dust of the ground, and breathed into his nostrils the breath of life; and he became a living being"; Genesis 3:7).

Though wounded and bloodied, Earth strikes back. Earth has its revenge. Earth does not passively acquiesce to Cain's attacks and stand by and watch his gory rampage proceed with impunity. On the contrary, Earth retaliates and "inflicts a curse" on Cain by "withholding its bounty" from this farmer-killer who now must roam the land unprotected and without security. Now that Earth, which once was Cain's friend and now is his enemy, will no longer "yield its strength" to meet his needs, Cain must leave his family home and become a nomadic "fugitive and wanderer" in search of the sustenance necessary for his survival. Genesis 4's intensely accelerative and explosive descriptions of Earth's *verb actions* in this account—the ground *cries out* over Abel's murder; it *opens its mouth* and *swallows* Abel's vital fluid; it *curses* Cain's perfidy and *refuses to give of its strength*—depict a profoundly agential life-form with its own interior psychology, purposeful behavior, and affective capacities for loss, anger, despair, abjection, and revenge.

The theme of the suffering Earth—a living organism with dispositions and intentions all its own—continues apace in other biblical texts as well. Throughout the writings of the prophets, human sin causes the land to become polluted, desolate, waste, and void. In the book of Jeremiah, for example, sin propels time to run backward in a manner analogous to a scene in a nightmare in which all of the action takes place in reverse. In Jeremiah, the everyday teleology of temporal existence collapses along with the life-giving energies of the biosphere itself. Like a cosmic clock whose hours and minutes hands are spinning backward, Earth returns to the amorphous state that characterized it at the dawn of creation in the opening verses of Genesis ("In the beginning God created the heavens and the Earth, and the Earth was formless and void"; Genesis 1:1). The prophet Jeremiah preaches that God's people, who have disobeyed God, will be taken captive by their Babylonian enemies and that Earth will resume its primitive formlessness. A well of sadness and loss will spring up in the land—the chosen people and the wider Earth community will suffer the agony of a world lost to artful practitioners of evil whose interior moral corruption causes creation's ecological despoilment.

> My anguish, my anguish! I writhe in pain! Oh, the walls of my heart! My heart is beating wildly; I cannot keep silent; for my people are foolish, they do not know me; they are stupid children, they have no understanding. They are skilled in doing evil, but do not know how to do good. I looked on the earth, and lo, it was waste and void; and to the heavens, and they had no light. I looked on the mountains, and lo, they were quaking, and all the hills moved to and fro. I looked, and lo, there was no one at all, and all the birds of the air had fled. I looked, and lo, the fruitful land was a desert. . . . For thus says the Lord: The whole land shall be a desolation, yet I will not make a full end. Because of this the earth shall mourn and the heavens above grow black; for

> I have spoken, I have purposed. I have not relented nor will I turn back. . . . For I heard a cry as of a woman in labor, anguish as of one bringing forth her first child, the cry of daughter Zion gasping for breath, stretching out her hands, "Woe is me! I am fainting before killers!" (Jeremiah 4:19, 22–28, 31)

In Earth's innermost life, its plight is to suffer distress and depredation—"the earth shall mourn and the heavens above grow black" (4:28)—and bear sorrowful witness to light being blotted out of the sky, earthquakes in the mountains, birds taking flight and fleeing certain destruction, and the verdant land becoming barren and dry. Like a woman in labor, the people hope to give birth to new life; but they are too late, and instead they are stalked by killers who lay waste to the body of the birthing mother and the body of the mourning planet itself.

The book of Hosea makes the same point. Because people have turned their backs on God's ways, the land will feel pain. And again, Earth is not a passive object but a fellow sufferer with its own consciousness and personhood. As did Jeremiah, Hosea laments that the chosen people's murderous intentions, just like Cain's, spell catastrophe for the land and its residents alike. Prophesying the world's destruction, Hosea bewails, "Hear the word of the Lord, O people of Israel; for the Lord has an indictment against the inhabitants of the land. There is no faithfulness or loyalty, and no knowledge of God in the land. Swearing, lying, and murder, and stealing and adultery break out; bloodshed follows bloodshed. Therefore the land mourns, and all who live in it languish; together with the wild animals, and the birds of the air, even the fish of the sea are perishing" (Hosea 4:1–3). Like a keening family after the death of a loved one, the beloved homeland bemoans its untimely death and the corollary destruction of its human, animal, and plant inhabitants. Grave planetary sorrow and the die-off of all living creatures is the by-product of human greed, an apocalyptic scenario not unlike our own time, when wave after wave of species' extinction is being fomented by extreme fossil-fuel extractive practices.

Similarly in the Christian scriptures, Paul writes that the whole Earth groans in Jeremian gestational pain as well. But now Paul, unlike the prophets, anticipates a future resolution of all beings' anguished longing. Human beings share in the natural world's painful waiting period—but together, both the sentient, injured Earth and distressed and hurting humankind anticipate some sort of final salvation. Paul writes,

> I consider that the sufferings of this present time are not worth comparing with the glory about to be revealed to us. For the creation waits with eager longing . . . in hope that the creation itself will be set free from its bondage to decay and will obtain the freedom of the glory of the children of God. We know that the whole creation has been groaning in labor pains until now; and not only the creation, but we ourselves, who have the first fruits of the Spirit, groan inwardly while we wait for adoption, the redemption of our bodies. (Romans 8:18–19, 21–23)

Paul is less clear than the prophets as to the cause of Earth's distress. He alludes to creation being "set free from its bondage to decay" (8:21), most likely a reference to the impact of human depravity on planetary well-being, in keeping with the Bible's general understanding of the causal connection between moral decay and nature's degeneration, as we saw in Genesis, Jeremiah, and Hosea. But what I find most striking about this passage is that the same maternal imagery that is deployed in Jeremiah is now used here. Drawing an analogy to a woman in labor, the created order of things, including, as Paul says, human beings as well, is in a protracted gestational state where painful groaning is an everyday reality. Metaphorically, Paul writes that creation is Mother Earth who yearns to get beyond the pain of childbirth to a time of deliverance from suffering. Once again, Earth is understood as a biodynamic personal entity, now in an analogy to a birthing mother, with its own cognitive faculties and emotional register. Pining for its deliverance from decay, longing for the day when it will be set free from its peculiar bondage, all of creation, and we too, moan inwardly for the time when the environmental squalor we have foisted on ourselves and our earthen home will be remediated.

Paul's groaning Terra Mater is a deep echo of the weeping land in Genesis, the anguished birth mother in Jeremiah, and the mournful ground in Hosea. Earth in the Bible, like Lovelock's "Gaia," is a living, feeling being who cries out and suffers injury from the depredation brought about by human malice. But why is this animist insight—the recognition of the common personhood of all things who suffer repeated injury—so crucial to our well-being on the planet? It is crucial because the existential awareness that we ourselves are not the only bearers of apperceptive suffering compels us to resituate ourselves—ontologically and ethically—in the wider personhood of Earth itself, who, like us, is a living being with emotion and purpose unto itself. It is crucial because this insight into our wider belonging to a living being far greater than ourselves compels us to reimagine ourselves as integral members of a cosmic body, a supreme organism, an all-encompassing life-form whose needs and requirements surpass our own and to whom we owe our ultimate loyalty and devotion. It is crucial because this recognition of Earth's vital essence forms the basis of more-than-human sacred kinship relationships and rituals wherein all beings are now regarded as sharing a common existence together as equal coparticipants in the web of life. And it is crucial because once we sense the longing of creation to be free from chronic suffering—once we sense nature's capacity to experience depredation in a manner similar to how we too experience loss and injury—then we will feel an inner drive to live our lives in harmony with all of God's creatures, all of whom, including ourselves, subside and flourish in Earth's loving embrace.[13] Or, as Lovelock puts it so succinctly, once we recognize Gaia as a "single living entity," we will then feel the "compulsive urge to belong to the commonwealth of all creatures which constitutes Gaia."[14]

## *Refreshment and Fragrance in the Hills*

It has been said that you cannot go home again. To be sure, the scent of sage is today lost to me when I revisit my early California home—a painful reminder to me of the ongoing depredation of Earth's ecosystems. In spite of this loss, however, I recently discovered the pleasures of wild sage again across a different landscape while traveling the Iberian Peninsula. In recent years, I have been fortunate to be able to enjoy everyday hints of this pungent blessing by spending summers at my in-laws' rustic townhouse in Palau-saverdera, a small agricultural village near the Mediterranean coastline in Catalonia, Spain, just south of the French border. The little town boasts two farmer's markets, an eleventh-century Romanesque church in honor of Saint Joan, and a small bar and cantina called Cal Pintor where the proprietor's mother prepares homemade paella. And beyond the borders of Palau-saverdera is a breathtaking, windswept national park that covers the easternmost point of the Pyrenees Mountains in northern Spain.

In the preserve, Audrey and I hike an arid, shrubland peninsula at the foot of the Pyrenees named the Cap de Creus (Cape of the crosses). The cape sits at the border between France and Spain and juts out into the Mediterranean Sea. The twentieth-century surrealist artist Salvador Dalí repeatedly painted the cape's sandy beaches and sea cliffs using a technique called *chiaroscuro* (light and dark), in which he emphasized the strong tonal contrasts that mark the shoreline's hidden caves, vertical precipices, and tortured rock formations. The Catalan historian Michael Eaude writes of the resonance Dalí felt with this stony landscape: "Cap de Creus has smooth rocks, wrinkled rocks, cleft rocks and rocks so sharp you cut your hands on them. It has rocks like fingers pointing in the sky, rocks with holes in them like so many of Dalí's figures and melting rocks like paste squeezed from a tube or like the limp, dissolving watches, the most famous icon in his painting."[15] Including its Daliesque coastline, the cape is a rugged terrain of spectacular vistas, vibrant flora and fauna, and strong gales called *tramuntana* (beyond or across the mountains) that sweep down from the Pyrenees Mountains in winter to the north across the cape and its coastline. On good days, the cape is surrounded by the clear turquoise water of the Mediterranean, and its mountain cliffs offer panoramic views of the sea, valleys, and small villages that dot the landscape below. But through the centuries, the icy blasts of the notorious *tramuntana* have ruined merchant vessels, pirate ships, and a few navy squadrons of some of the great Western maritime powers, including the Phoenicians, Greeks, Romans, and Arabs.[16]

According to legend, it is the cape's violent weather that gave the peninsula and surrounding sea its name. Its jagged shoreline and rocky seafloor is the final resting place for wooden vessels that sank due to the Pyrenees's gale-force *tramuntana*. In stormy conditions, these boats sank straight down onto the seafloor, with their vertical masts and horizontal crosspieces extending upward through the water—creating, as it were, a cemetery of underwater "crosses." The Cap de Creus is a graveyard of shipwrecks. But today, the cape, still rock strewn and forbidding, is also a beautiful chaparral of juniper and ash trees; an aromatic nursery of wild rosemary, lavender, thyme, heather, and rock roses; an arid home for many species of toads, lizards, and geckos; and an aerial habitat for the gulls, terns, falcons, and eagles that soar though the sky overhead.

Archaeologically, the Cap de Creus is a union of ancient and medieval settlements. It is dotted with Neolithic *dolmens*, small tables of standing rocks, with large, flat roof stones at their tops, that were used for temporary shelters, or perhaps burial sites, for hunter-gatherer peoples millennia ago. These prehistoric sites are nestled amid the ruins of once-grand olive groves and grape orchards that covered the cape roughly a thousand years ago. These groves and orchards were cultivated by Benedictine monks in the ninth and tenth centuries CE for the production of oil and wine. Hiking in the area, one can still see the traces of the terraced gardens and stone enclosures that still grace the hillsides. As well, the Benedictines built a spectacular Romanesque monastery called Sant Pere de Rodes (Saint Peter of Rhodes), now a restored ruin and active pilgrimage site, perched high on a ridge of the Pyrenees overlooking the Mediterranean Sea.

The monastery is complemented by a nearby seventeenth-century mountain chapel, the Sant Onofre (Saint Onuphrius) Hermitage, seemingly suspended in midair across the splintered rocks of the Pyrenees foothills. The small chapel is named after an Egyptian anchorite, Onuphrius, from the fourth or fifth century CE who imitated the austere, reclusive life of John the Baptist—much as John Muir did in Yosemite some fourteen hundred years later. According to legend, Onuphrius lived by himself as a desert hermit in remote areas of Upper Egypt. He is often depicted as a wild man with flowing white hair, a loincloth of leaves, and a walking staff. Onuphrius the renunciate relied on the natural gifts of the desert for his subsistence, crediting to God's providence "a palm tree [that] grew next to his cell which brought forth dates in abundance and a spring of water [that] began to flow there."[17] Today this well-preserved but remote chapel in the eastern Pyrenees is cantilevered over a rocky outcrop and is said to be chosen, in honor of Onuphrius, for its proximity to a freshwater underground spring in the midst of the promontory's harsh environment. The desert fountain at Palau-saverdera's Onuphrius chapel, in keeping with its monastic namesake, is a daily witness to nature's generous abundance.

Throughout Christian history, many temples and churches were commissioned to be built nearby "holy wells" or "sacred springs." In the British Isles, a variety of churches, often continuing pre-Christian sacred water veneration, were founded at the site of underground springs, including Winchester Cathedral in Hampshire, England; St. Patrick Cathedral in Dublin, Ireland; and Glasgow Cathedral in Scotland. In modern-day Nazareth, Israel, the Greek Orthodox Church of the Annunciation was established at "Mary's Well," the still-surviving watering hole where the Angel Gabriel announced to Jesus' mother that she would give birth to God's son. At Montserrat (serrated mountain), just outside of Barcelona, Spain, the Benedictine abbey church and a series of nearby mountain chapels were built next to a running stream in a sacred grotto where the small blackened statue of Mary I noted in the preface was first said to be discovered and is now visited by throngs of pilgrims.[18] Across time, people have been awestruck at the wondrous gratuity of nature to give continually of itself all that is necessary for a flourishing life, including neverending liquid refreshment, and they have built small sanctuaries and great shrines as testimonies to the benevolent power of such places. Bubbling springs create oases of verdant plant and animal life and awaken a sense of numinous powers—sometimes called water sprites or water nymphs in traditional mythology—present at the site of such magical places. Whatever the presence of such beings might be at Saint Onuphrius, this sacred grove—a rich habitat nourished by the underground steam with lush foliage and thriving birdlife—provides a welcome respite from the Mediterranean sun, howling winds, and steep rock faces that define the rugged setting of the cape.

Hands down, the hermitage spring gushes forth the sweetest water I have ever tasted. Hot, sweaty, parched, exhausted—Audrey and I tumble down the scree to the chapel fountain brimming with excitement. Filling up our water bottles, washing our faces in the spray, soaking our shirts in the cold stream, we take deep draughts of liquescent grace from the natural seep, laughing at the overflowing extravagance of such a miracle within the confines of the cape. The water tastes so good—it has a fresh and pure flavor with a touch of honeyed sweetness and a slight mineral tang to it—and we marvel at our good fortune that in spite of the promontory's broiling heat and climatic vagaries, we can enjoy an interval of peace and relief at the bubbling wonder of the Saint Onuphrius spring. This hallowed watering hole is a living testimony to creation's boundless, ever-flowing largesse, like Augustine's eternal fountain or Muir's purple Sequoia juice, that continually wells up from within God's bosom to renew and heal every denizen of the good Earth.

My and Audrey's hikes in these mountains—a walk back in time that traverses centuries, even millennia—often begin and end at the site of this spring. But this is not the only sensual, libidinal pleasure I experience on the peninsula. In addition to fresh water, the cape offers wave after wave of sagey fragrance that washes over the ragged surface of the rockscape. On our rambles, I trek the cape intentionally and prayerfully, with my head slightly lowered, my gait deliberate, and my mind focused on the rhythm of my breath—in and out, in and out, in and out—as I take in the intoxicating zest of the vegetation that surrounds us. Breathing fully and inhaling deeply, I am gifted with a heady bouquet that refreshes my body and uplifts my spirits. For me, our cape walks are a contemplative rite in which walking, seeing, feeling, hearing, tasting, and smelling merge into a spicy, fleshly encounter with divine presence. As well, they return me to my nighttime sleep and dream patterns in the warm embrace of the sagebrush hills of my boyhood. Now time runs backward, not toward the chaos of biblical apocalypticism but toward the aromatic balm, carried on nighttime breezes, that floated gently through my California bedroom many years ago.

On the Cap de Creus, the sage is plentiful, alive, and fragrant. As a national park, the cape is not to be disturbed, and so its many endowments—including its gifts of spring water and wild herbage—are proffered for pilgrims like me who hike its environs to find peace, to find strength, to find hope, to find God. Through our addiction to fossil fuels in particular and unsustainable consumption in general, our species has wreaked havoc on our planet. Like Abel's blood at the hands of his murderous brother, Cain, the blood of our many victims cries out from the ground. As is lamented in Jeremiah, Hosea, and Romans, as we saw, Earth mourns, the animals and plants continue to perish in astonishing numbers, and the underresourced and dispossessed among us languish in squalor and despair. We have laid waste to Earth, plundered its abundance, stripped bare its many gifts, and left it susceptible to ongoing injury and exploitation. Poverty stricken, vulnerable, and seemingly helpless, all of creation continues to groan in labor pains for a future deliverance about which we know very little.

In the face of this desolation, are there inklings of hope? For me, there are still occasional droplets of grace—an ancient watering hole nestled within the aromatic shrublands of northern Spain—within the tempest of encroaching ecocide. Even now, there are flickers of promise amid the darkness and despair of a world set afire by the extraction industries—a world that is *warming* or, to put this point more clearly, a world that is *dying* from our continuous burning of oil, gas, and coal. My hope in living with, and perhaps at times in overcoming, this suffering that our global abuse has wrought is to ritually rediscover my innermost and most heartfelt *sense of belonging* with the natural world—in my case, a sense of belonging made possible by occasional hikes through the spring-fed sage lands of the Cap de Creus—and thereby find rest and security, and renewed energy, in the great plentitude of creation that God has made for our common life and common joy.

## *A Tramp for God*

Hiking the easternmost Pyrenees is a root metaphor for my Christian animist journey through a world saturated with grace but riven by humankind's assaults on the integrity of this blessed world. I recently lived this metaphor once more with Audrey in another part of Spain: the El Camino de Santiago (The way of Saint James) pilgrimage route in Galicia, Spain's westernmost province on the Atlantic coast, just opposite the Mediterranean coastline where we hiked across the Cap de Creus. The origins of this trail date to the tenth century CE. Historically, along with Rome and Jerusalem, the Camino de Santiago was one of the three great pilgrimage routes of Christian Europe.[19] Whereas the cape headland on the Mediterranean is rocky and forbidding, much of the Camino in Galicia winds through a leafy bioregion of subsistence farms, small villages, and lush forest and countryside. The provincial name, Galicia, is Gaelic in origin, and the Celtic look and feel of this northwestern Spanish landscape is a lot like Ireland and Scotland, with strong genetic and cultural roots tying the different regions together as well.

Today, hikers along this ancient route often stay in the same stone-built pilgrim or convent hostels wayfarers used during their own pilgrimages hundreds of years ago. There is no cost to walking the Camino, and while anyone can hike the route, most of the travelers I met saw themselves as spiritual seekers in quest of some sort of purpose and meaning in their lives. The hikers I encountered along the way—an atheist Canadian Catholic priest leading a confraternity, a Spanish man distraught over the breakup of his marriage, a middle-aged woman from Ireland who offered to Audrey a panty liner for wrapping her blistered feet—were seeking to be more than backpacking tourists but, rather, intentional pilgrims of one kind or another, whether they self-identified as conventionally religious or not.

Contemporary pilgrims on the Camino carry with them a *credencial* (pilgrim passport) that takes a rubber stamp from local bars, hostels, and churches to prove that they have walked the route. They follow signposts marked with clamshell decals and carvings that symbolize the martyrdom of the apostle James—whose body, according to one legend, was thrown into the sea, after an early evangelistic mission in Iberia, only to be recovered for burial fully intact several days later, but now covered with clamshells.[20] These days pilgrims often attach a clamshell to their backpacks in honor of the route's patronym. The Camino covers thousands of kilometers and many different pathways across half a dozen European countries. But if a traveler walks at least the last one hundred kilometers, he or she is awarded a *compostela* (certificate of completion) from the cathedral in Galicia's capital city, Santiago de Compostela (Saint James of the Field of Stars), which is also the end point of the pilgrimage route. Often, as in my case, one's first name is rewritten in Latin on the *compostela* at the end of the journey (my name was changed from Mark to "Marcum").[21] The Camino, then, not only provides pilgrims with a new identity but also offers them opportunities for reorienting their lives toward whatever they find to be worth living for. In medieval times and today during particular holy years, a *compostela* also signifies that a pilgrim is the recipient of a "plenary indulgence"—the Catholic church's postmortem assurance to the faithful that their sins are completely forgiven and that no time in purgatory will be required accordingly. In the present, whatever one thinks about indulgences, the Camino provides other tangible opportunities for existential reorientation through ritual activity: the Cruz de Ferro (Cross of iron), to take one abiding example, is a makeshift pole with a cross on top where wayfarers can leave, for example, a pack of cigarettes or a photo of a recently passed loved one, as tangible signs of their own journeys to healthy living or as marking the beginning, or the end, of a period of mourning.[22]

For me, I travel this route in order to find something within myself—a sense of calling or vocation—that will be awakened by the beauty and the challenge of the Camino's rustic environs. My aim is to nurture the occult relationship between the "inner landscape" of my spiritual identity and the "outer landscape" of the pastoral setting of the Camino—to correlate the private camino of my heart with the public Camino of the pilgrim's way. As Muir writes, "I only went out for a walk and finally concluded to stay out till sundown, for going out, I found, was really going in."[23] Or as John Brierley, my favorite Camino travel-guide writer, puts it,

> A pilgrim travels on two paths simultaneously and must pay attention to both. When we place ourselves on the pilgrim path we sow the intention to stretch and expand soul consciousness so that we can lift ourselves out of the mundane in order to journey back to God or whatever name we give to that nameless Source from whence we came.
>
> When asked to describe a personal experience of the sacred, an overwhelming majority refer to a time alone in nature, "A sunrise over the sea; animal tracks in fresh powder snow; a walk under the full moon." . . . [These experiences act] as a reflection of a larger perspective, a distant memory of something holy—something bigger than, and yet part of, us. This is where the camino can provide such a powerful reminder of the sacred in our lives and the desire to reclaim our spiritual inheritance.[24]

In my theology, the so-called mundane world and the heavenly world are one and the same. Pace Brierley, the traveler does not need to be lifted "out of the mundane in order to journey back to God" because the mundane *is* always-already permeated with intimations of divinity everywhere. To trade on spatial tropes, God is "over there" or "in here," not "out there" or "beyond." But Brierley's point is otherwise well taken: for the religious pilgrim, the *real* Camino oscillates between the outward pilgrimage to Santiago de Compostela, wherein traces of God can be found all along the way, and the inner pilgrimage to one's deepest self, where signs of divine presence are equally discoverable. Rambling in John Muir's Yosemite highlands as a young man, hiking through the tropical forests of Monteverde in midlife, and, lately, walking the well-worn paths that mark the Spanish Pyrenees and the Camino have taught me that the physical act of simply, but mindfully, putting one foot in front of the other gives me a sense of spiritual progress amid the challenges and upheavals of everyday existence. Grabbing my staff and clamshell, becoming, in that week that Audrey and I hiked the Camino, a tramp for God, a Christian itinerant, a Gospel bum, to paraphrase Jack Kerouac,[25] the Camino became an opportunity for me to recharge the emotional wellsprings that energize my own daily peregrinations through a beautiful world pockmarked with sorrow and heartache. I reminded myself again that life is a journey. Existence is a sojourn. To be human is to travel a pathway of indeterminate origin and uncertain destination, sometimes with a feeling of calm purpose and sometimes with a sense of irrecoverable loss that is beyond repair.[26]

Here are my spotty journal entries from September 2011, when Audrey and I spent a week on the Camino—hopeful for spiritual renewal but ill prepared, as we quickly discovered, for hiking long distances with poor footwear:

> 1st day of Camino. Writing this in a 1 star hotel in Portomarin in No. Spain en route to Santiago de Compostela. With Audrey, today was a magical trek through small villages and pasturelands. Nothing can compare to blue sky, green meadows, and an occasional car or dog—or fellow pilgrim—entering the path alongside us. I was esp. enlivened by the gustatory treats—sweet and savory—afforded hikers both at roadside tabernas—coffee, wine, potato omelets, house pudding—but most of all as part of Earth's fruits—wild blackberries and raspberries, tart apples, and cold mountain water from specially marked fountains. My and Audrey's feet hurt like a thousand demons—protracted foot rubs helped—as we sank into bed listening to the carillon music of the town church nearby.
>
> 2d day of Camino. While day began well—brilliant sunshine and easy hiking—the pain in my and Audrey's feet took over the pleasure of the countryside. At the peak of Sierra Ligonde I felt some relief

because of a light mist obscuring the dancing sun. But blisters began to spread from toe to toe. In Palas de Rei that evening we spent a couple of hours rubbing each other's broken feet while Audrey lanced my blisters—and I hers. We drank a bottle of rural estate tinto and devoured a traditional stew-like soup just before bedtime. I fear the prospect of finishing the Camino this time is not good.

3rd day of Camino. Feeling refreshed, we powered through 15 km in 4 hours. After lunch, however, we faltered as our blister problems—and the threat of Audrey losing a toenail—continued. The beauty of Spain's farms and hamlets remained: sun-dappled woodlands with oak and eucalyptus, quiet streams crossed by medieval foot-bridges, flute-like bird calls, barking dogs, even the occasional pilgrim with cell phone beeping. The romance of the Camino is poignant: "Maria, you are my camino now," read one graffitied wall in a highway underpass. I am deeply troubled by the martial tone of the journey with "crusaders' crosses" everywhere. But a quiet morning at the 13th c. Romanesque church in Lobreiro—replete with a stone altar, Mary dressed in a wedding gown in a side niche, and a spectacular wall painting of Jesus' passion—made the long hike worth it, in spite of my and Audrey's screaming feet. I doubt we can continue—Audrey is such a love, I hate to see her suffer—but every morning brings new hope.

4th day of Camino. Our feet have recovered somewhat—or we are becoming inured to the pain—as we traveled through woodlands and tiny villages. Found ripe figs and more raspberries to supplement roadside taverns and cafes. When I saw the fluid spurt from Audrey's punctured blisters like a geyser I was sure we were quitting the camino. But the new morning brought new hope. Now—if possible—to Santiago.

5th day of Camino. Walnuts and acorns peppered our path—borders of rosemary and mint as well—as we walked through oak and beech groves toward Santiago. Our feet betrayed us at every step. But the goal of reaching the cathedral buoyed our spirits. Sightseeing "pilgrims" and tour buses crowded the way, but when we arrived in Santiago—skipping lunch—resting in the church, getting our (different!) *compostelas*, and arriving by bus back to Sarria that night, we slept the warm sleep of secure children.

The Hinduism scholar David Haberman writes that there are two types of pilgrimage: linear movement using a structured travel plan in which one's destination toward a center or end point is fixed beforehand, and open-ended meandering through a web of possibilities with no predetermined center or end point. He proposes a "simple typology of pilgrimages, which compare pilgrimages that value a center with those that devalue a center. The first type involves a linear journey and has a clear destination, a center; the second type involves a circular journey that resists the center and has no clear destination. The implications of the differences between these two types of journey are significant."[27] In the first case, the pilgrim is a purposeful devotee with an itinerary he or she follows to reach a transcendent goal, while in the second, he or she is a nomad who aimlessly drifts across a serpentine landscape always ready and open to traces of the everyday sacred. To paraphrase a simple formula, for the devotee, the destination is the journey; for the nomad, the journey is the destination. In the first instance, obstacles to reaching the center or the end are challenges that must be overcome, while in the second, they are occasions for enticing detours and errant wandering that are themselves the point of the journey, not the preset finish line.

But, of course, many pilgrimages are a mix of winding travel and following a fixed route. In this combined effort, the religious pilgrim is both awake to the ubiquitous sacred embodied in all things and mindful of the eventual goal, toward which he or she aspires, where divinity is also powerfully present. My experience of the ancient Camino route suggests both modalities were at play in my and Audrey's travels. We were walking a line, but we often got lost; and our wayward excursions were often welcome respites from our goal-oriented itinerary. I recall Audrey and I spending part of one day enjoying a long lunch on a small bridge overlooking a brook because I had confused, in my fatigue, the clamshell Camino way sign with a similar trail marker for another route. My journal entries remind me that our blistered feet—Audrey lost two toe nails along the way—were a painful auger of our likely inability to finish the Camino. But there was no bitterness about not reaching our stated purpose; we had a finishing point in mind, but our goal was also goalless; and if we did not reach the faraway cathedral, every (often painful) step was still worth the effort (or so it seemed at the time). Our hope was to reach Santiago de Compostela and receive a *compostela*, if not a plenary indulgence, for our efforts—a testimony to our weeklong agon. But equally, if not more important than the goal itself, were our encounters with the scattered graces we came upon all along the way: fresh spring water, wild figs, local wines of the *terroir*, generous fellow travelers, and wayside altars, shrines, and churches on the path that reminded us of the generosity of the Galician people—indeed, a whole society's common commitment to hospitality and the well-being of pilgrims such as ourselves. But of course, in the end, after reaching Santiago, we simply got on a return bus back to the village of Sarria, from whence we had begun our trek. For me, the Camino was both a line and a circle, a focused destination and a circuitous wandering, a labyrinth with a fixed center and a maze with no necessary end point.

Beyond my journal entries, I am mostly reminded today of the rich *sensory* experiences I felt along the trail. Wayside massage parlors that promise to restore pilgrims' broken feet. Ribbons of fragrant wood smoke rising over rolling vineyards in the distance. Cave-like grottoes where flickering candles illuminate grotesque medieval frescoes of Christ's passion. Village squares redolent with the briny smell of *pulpo* (octopus) frying in giant cast-iron pots. Roadside cafés that open at sunrise to serve travelers hot chocolate and *churros* (fried dough covered in sugar and cinnamon). Nesting white storks perched high on the gabled rooftop of a flagstone church. Disturbing crusaders' cross altars perched high above, reminding us of the Camino's brutal history as a symbol of the Santiago Matamoros (Saint James the Moor-slayer) legend, in which the apostle James returned to Spain in the ninth century CE to slaughter Muslims in a bloody battle. And a thin young man with bowed head and bare feet, carrying a shepherd's crook and water-filled gourd as a canteen, clothed in a tattered robe belted with a rope and imitating the stride, the look, and the piety of a bygone era on the trail.

## *The Death of God*

Near the end of the Camino, Audrey and I arrived at the last scenic high point before entering Santiago de Compostela. Monte del Gozo (Mount joy) is a windswept overlook on the outskirts of the city, with the spires of the Santiago cathedral rising in the distance. It was filled with sightseers and backpackers like ourselves the morning we arrived, many eating at outdoor tables. Mourning doves and feral pigeons were everywhere eating leftover crumbs of food as pilgrims refreshed themselves one last time before arriving with the crowds at the cathedral, only half a kilometer away. The doves were familiar in their buffy gray- and cream-colored plumage; the pigeons sparkled in silver feathers with iridescent green and purple highlights. The one group produced their mournful cooing calls, while the other made soft gurgling sounds. Commonplace and insignificant, the doves and pigeons bounced and fluttered together across the lookout as they picked at the table scraps left behind.

It is easy to suppose that ordinary birds such as doves and pigeons will always be with us. Mourning doves are everywhere, and city pigeons (rock doves) even more so. We assume that their sheer numbers will protect them against depredation—and that their total extinction is unimaginable.

But the extirpation of the passenger pigeon in North America is a cautionary tale about the seeming impossibility of wiping out whole species of overabundant birdlife. Prior to the twentieth century, the passenger pigeon numbered in the multiple billions. Certainly in America, and perhaps globally, the species had lived in abundance and in harmony with its wooded surroundings for tens of thousands of years. Flying overhead, enormous flocks of passenger pigeons were compared to aerial waterfalls or massive cloud formations that blotted out the sun's light, sometimes for days, over whole towns or farmers' fields. Onlookers said that the deafening, rumbling sound of the birds' wings, beating the air above, sounded like a freight train passing through. But widespread commercial hunting of wild pigeons—trapped in nets, baited with poison, blasted with shotgun spray, burned out of their nesting sites—led to the end of the whole species. Jean-François Millet's 1874 painting *Bird's-Nesters* is a hauntingly graphic portrayal of French peasant boys actively blinding thousands of passenger pigeons with torchlight—and then wildly clubbing them to death under the dome of a close-to-the-ground, sickly greenish sky. The wild frenzy of the birds' beating wings, as many drop to the ground, is a blur of blood and chaos that is almost impossible to take in, even in a painting that is more than a century old.[28]

Heralded as the "ecosystem engineer of Eastern North American forests," the passenger pigeon today is recognized as having performed an outsized service in promoting biological diversity in precontact and post-contact early America.[29] As seed eaters and foragers, swarms of pigeons would blow through forested areas dispersing large seeds that they had consumed in flight, on the one hand, and smaller seeds in their biologically rich guano, on the other. The current decline of North American woodland ecosystems is now, in part, attributed to the persecution of this wild pigeon. As a cheap source of protein, the pigeon lost its status as a symbol of nature's seemingly endless abundance and became simply one more market commodity to be exploited and sold, even to the point of extinction. The last passenger pigeon, Martha, died in the Cincinnati Zoo, Ohio, in 1914. She was estimated to be twenty-nine years old.[30]

*If God is a bird, then what is the sense of loss, not only biologically but religiously as well, when innumerable members of a particular avian species (e.g., the passenger pigeon) are senselessly slaughtered?*[31] In addition to the role pigeons (and all other birds) play in promoting healthy ecosystems, what about the role they play in ensuring the *spiritual* health of human beings? The American poet W. S. Merwin speaks of his gratitude for the countless creatures in our midst but says that their dying in great numbers all around us evacuates our emotional lives and erodes our spiritual selves. He writes,

> with the animals dying around us
> taking our feelings we are saying thank you
> with the forests falling faster than the minutes
> of our lives we are saying thank you
> with the words going out like cells of a brain
> with the cities growing over us
> we are saying thank you faster and faster
> with nobody listening we are saying thank you
> thank you we are saying and waving
> dark though it is.[32]

Merwin says that with the animals dying and the forests falling faster than the passage of time itself, we are thankful even while waving good-bye to our creaturely and arboreal friends. When we drive into oblivion the animals and trees who encircle our everyday lives, all we can do is wave good-bye and say "thank you faster and faster," even though no one is listening and even while we abandon to extinction the living beings—the traces of divinity—who are now "taking our feelings" with them in their passing. If God is a bird, and if we persecuted to death a primary member of the family of dovey pigeons—namely, the passenger pigeon, a cousin of the beaked deity, the Holy Spirit, who graced Jesus' baptism—then what happens to us and what happens to God and what happens to God's relationship with us when we have killed off such a bird? When we extirpated the passenger pigeon and when we ensure the ongoing loss of the "animals dying around us [who are] taking our feelings," then what happens to our *felt* experience of God's habitation in the world when these particular winged enfleshments of divinity are no longer with us? The affective presence of God's Spirit in daily life is radically weakened when such divine embodiments (we could say *enfeatherments*) are annihilated in the multiple billions of creatures, as happened in the case of the passenger pigeon. My suggestion is that the tragedy of the pigeon one hundred years ago is the tragedy of God in our time and place as well: the mass extinction of this wild bird blots out the felt presence of divinity among us. The cruel loss of the passenger pigeon further deepens our sense of God's trauma, suffering, and absence in our own time—a time scarred by the demise of the countless avian cousins of Jesus' sacred baptismal bird, such as Martha, perched stiff and glassy-eyed in a glass case at the Smithsonian National Museum of Natural History in Washington, DC.[33]

The Christian belief in the incarnation of God in Jesus and the embodiment of the Holy Spirit in the baptismal dove should actually mean something in practical, spiritual terms, not simply serve as points of theological doctrine. If God shows Godself to us in the gift of the sacred dove at the time of Jesus' immersion by John, then all doves and, by extension, all creatures are sacred beings. Thus, to inflict needless pain and suffering on any one such being—much less one whole species of being, such as the passenger pigeon—is to inflict needless pain and suffering onto God. If we do chronic and lethal harm to the beings in our kinship circle, do we not place God in serious jeopardy, even render God vulnerable to death as well? Matthew Eaton makes this same correlation in his recent Ph.D. dissertation. He says it is time to consider whether God's kenotic outpouring of Godself into worldly flesh is an exercise in divine misery and destitution because the world itself is dying. Eaton writes,

> The urgency of our era necessitates such risks in order to explore theologies willing to call ecological violence into question, even at the cost of partnering with frameworks heretofore viewed as heterodox. . . . [These frameworks specify that] the role humans play in the violence of ecocide necessarily correlates, beyond the death of Earth, [with] the possibility of deicide, the death of God. If such is the case, the anthropogenic death of God would not lead to life . . . but rather to an eternal trauma within the religious ecology in which all things live, and move, and have their being.[34]

*As the world goes, so goes God.* To torture and torment the successor bird to the dovey pigeon who corporealized God's Spirit at the River Jordan is to torture and torment God too. Exterminating the passenger pigeon—one of the many avian faces of God in our midst—hopelessly deadens God's nearness and intimacy to each of us, running the risk of effacing God in our time as well.

Like Paul's groaning Terra Mater, the cry of the Earth today is a deep and continual lament. It is the burning question to each of us: Will it ever be possible for us to realize that by wiping out whole species of avian divinity we are destroying the signs of God's loving and abiding immediacy to each of us as well? Will it ever be possible for us to comprehend that "when the animals dying around us [are] taking our feelings" with them, we are losing our affective capacities to sense the sacred within the everyday? No longer flocking overhead in thunderous clouds of wonder and power, the extinction of the passenger pigeon erases even further hints of divine immanence in commonplace existence. We claim to want to see and feel God within the ordinary world. But God's promises of the Spirit's ongoing presence in our midst—promises realized by each of us in our relations with the winged companions that accompany our comings and goings—are vitiated by the injury and death suffered by these enchanted beings who once took flight in our world and are now gone for eternity. *Our forbears executed God's innocent son at Calvary in a paroxysm of rage and violence; we do the same by crucifying God's winged Spirit on the Earth through market forces and habitat destruction.* God is crucified afresh when we lay waste to the carnal presence of God on Earth. The paschal trauma of the cross is daily reactualized through our regular assaults on the good creation God has made. The Earth has become *cruciform*: the scars of Golgotha are everywhere. Jesus' crucifixion wounds are now reopened as the whole Earth bears the marks of eco-catastrophe. The systemic murder of the passenger pigeon and the continuing collapse of species and ecosystems—myriads of creatures and habitats that were once thriving among us in abundance and now eradicated forever—are the death of God in our time and place as well.

When I encountered the busy doves and pigeons at Mount Joy, I thought to myself, "If in Christian witness, God is a dove, then are not these doves and pigeons around me embodiments of divinity as well, even though, as ordinary scavengers who seem to count for nothing in the great scheme of things, my first instinct is not to recognize them as such?" The world is limned with traces of God's presence, but these traces remain veiled for most of us, even those of us who desire to follow signs of the sacred in everyday life.

This phenomenon of divine hiddenness (*deus absconditus*) is not only the case today but also characterized the time, arguably, when divinity, at least for the Christian community, was most obvious in the world during the lifetime of Jesus of Nazareth. Yet one of the strangest themes of the Gospels is that Jesus was often unrecognizable even by those who were closest to him. In the frightening story, for example, of the disciples' storm-tossed journey on the Sea of Galilee—the story Muir reworked to narrate a visit by the water ouzel to his own distressed sailing expedition in Alaska—Jesus journeys out across the whitecaps to comfort his followers, but they mistake him for a phantom. "But when [the disciples] saw him walking on the sea, they thought it was a ghost and cried out" (Mark 6:49). It may be that the stress of impending shipwreck blinded Jesus' sailing party from recognizing their leader. But after he had previously stilled the raging waves in this same body of water in Mark 4:35–41, it is peculiar that his followers failed to identify him in almost the same set of circumstances sometime later. Two other Gospels make the same point. In Luke 24:13–35, two of Jesus' followers, Cleopas and an unnamed companion, are walking with Jesus, after his resurrection, on the road to the village of Emmaus, and while they seem to enjoy a lengthy hermeneutical conversation—"beginning with Moses and all the prophets, [Jesus] interpreted to them the things about himself in all the scriptures" (Luke 24:27)—Jesus' identity remains obscured to his two friends. Likewise, in John's Gospel, Mary Magdalene stands directly in front of Jesus in his tomb early on the morning of his resurrection, but she does not recognize him. Crying, she queries two angels with her in the crypt about where Jesus' body has been taken, and then "when she had said this, she turned around and saw Jesus standing there, but she did not know it was Jesus" (20:14). To be sure, Mary was upset, and the light in the burial cave might have been dim; but that evening, Jesus' other followers *did* recognize him—even though they were holed up at night in a locked (and perhaps darkened?) house for fear of persecution.

My point is that the Gospels often portray Jesus' identity as masked or hidden, even from his closest allies. Startlingly, at different times throughout the Gospels, the disciples, Cleopas and his companion, and Mary Magdalene herself were in total ignorance about who it was next to or standing directly in front of them. How is it possible that Jesus' closest followers did not recognize his identity *at all* even when they were up close and personal to him in these direct, often face-to-face encounters? I am not sure of the answer to this question—perhaps fear or trauma or his (slightly altered?) postresurrection visage played a role in preventing Jesus from recognition by his companions—but whatever the explanation, I find this phenomenon instructive concerning the difficulty, if not the impossibility, of recognizing divinity in the world.[35] If Jesus as God-in-the-flesh is unknown by his friends, then could it not follow that other enfleshments of God in everyday life, such as the flocks of doves and pigeons at Mount Joy, just like the erstwhile scores of (now extinct) passenger pigeons in eastern America, are also veiled, even invisible to human perception (*ales deus absconditus*)?

John's Gospel says that God becoming carnal in Jesus is the entry of divine *light* into the world ("And the Word became flesh and lived among us, and we have seen his glory"; 1:14). It continues that as God gifts the world with Jesus, God also blesses Jesus, at the time of his baptism by John the Baptist, with the Spirit in the form of a dove ("And John testified, I saw the Spirit descending from heaven like a dove, and it remained on him"; 1:32). And finally, John writes that at the time of Jesus' departure,

Jesus will give the gift of the Spirit as his emissary to the whole world—even as God gifted the world with Jesus prior to the Spirit's full arrival ("And I will ask the father and he will give you another advocate, to be with you forever. This is the Spirit of truth. . . . You know him, because he lives with you, and he will be among you"; 14:16, 17). This, then, is the core of my biblical argument for Christian animism: as God enters the world in Jesus, now Jesus, blessed by the Spirit, infuses the world with the continual offering of the Spirit—the selfsame Spirit, in John's Gospel, who lives in and among all things forever. As God beatified the world through the reality of Jesus' existence *at one time*, so also is God, through the gift of the Spirit, *continuing* to endow the world with abundant blessing today. Herein the world is redeemed. Herein all things are suffused with grace. Herein the whole Earth is full of God's glory, as Isaiah sings (6:3). In this extravagant benefaction, the world is a Spirit-imbued habitat of illimitable benediction. It is God's tabernacle wherein all things are sacred—including, and especially so, the consecrated doves and holy pigeons of Mount Joy, who reminded me during the Camino, and again today, of the Spirit's (onetime and ongoing) birdy incarnation at the event of Jesus' baptism.

Ironically, however, while biblical faith witnesses to a world filled with intimations of divine presence, it is oftentimes the Christian religion *itself* that is the primary obstacle to recognizing the godliness of winged creatures, indeed, the sanctity of all beings within everyday existence. If we think of Jesus' presence, now mediated by the Spirit, as the light that illuminates the reality of worldly grace ("And Jesus said, 'I am the light of the world'"; John 8:12), then we can think of Christian faith as a reflection of that christo-pneumatic light. Perhaps, to make the analogy clear, we should call Christianity the *shadow puppet* of the light: it itself is not the light, but it takes the light refracted within all creation and contours it into shapes that are meaningful to human perception. While Christianity itself is not the light, the light is all around us—the trill of the wood thrush on a summer afternoon, the aroma of wild sage drifting across the desert scree—and it sometimes helps to attach words and images to this light so that we can focus on its meaning and importance. But Christianity also gets in the way of the light. First, it deflects its rays from reaching our deepest selves through world-denying doctrines that shield us from the light around and within each of us. Second, it sometimes blocks out the light entirely by actively denying the presence of light in the world—what has been called "common grace" or a "sense of divinity" in a variety of Christian traditions.[36] And third, Christianity at times claims to be the light itself—for example, Pope Paul VI in *Dei Verbum* says that the church's teachings are a divinely revealed deposit of truth[37]—when, in reality, it is only one of the many shadow boxes that reflect the light. *No religion, Christian or otherwise, is the light.* Rather, the light is always-already here and everywhere, reflected and refracted in all things, and it far outshines all of its self-appointed human administrators and institutions, ecclesiastical or otherwise. In the end, the light is itself the light; it permeates and illuminates every corner of creation, and no human attempt to curate its display can alternately obscure or define its elucidating presence, its flickering continuity, its flashing brilliance.

## *God on the Wing*

Many of us today are enduring the shock and profound loss of living in the period of what some people are calling "the anthropocene," "the end of nature," or the time of the "sixth great extinction."[38] Due to anthropogenic climate change, the contemporary loss of plants and animals appears to be ten thousand times greater than naturally occurring extinction rates, as we saw earlier, with tens of thousands of species lost annually. The hopelessness that characterizes our looming dread about fracturing biodiversity—Greenland lost a trillion tons of sea ice in just the past four years alone due to global warming[39]—leaves many of us unsteady and fearful about the future. At times, historical ascetic Christianity has contributed to this feeling of doom by making war against the natural world and the flesh and by denying to God a place in the living world, rendering the biosphere a dead place, an empty place, a godless place.[40] Personally, I feel this loss profoundly, and the latest North American presidential election, and the role religion played in that election, has left me with even more despair: U.S. president Donald Trump has called climate change a hoax,[41] while 82 percent of white American evangelicals voted for this climate-denying presidential candidate.[42]

But Christian Trumpism runs aground on the shoals of the biblical animist belief in the living goodness of all inhabitants of sacred Earth. At the end of the world, the hope for a renewal of what William James calls "healthy minded" as opposed to "sick soul" religion rests on its practitioners regaining a profound sense of the fleshy, this-worldly identity of the sacred.[43] The philosopher Gayatri Spivak writes that "animist liberation theologies" could be the next political innovation on the horizon following progressive Christianity's recent turn to liberation and ecology.[44] If Spivak is right, then animism is ready to be recovered as the lost treasure of healthy religious belief—and the ground tone of its revival today. In the case of Christianity and *in* spite of and *to* spite its erstwhile anthropocentric chauvinism, God is not, in a biblical neoanimist worldview, a bodiless heavenly being divorced from the material world but is manifested instead as the winged Spirit who protectively alights on Jesus at the time of his baptism and thereby infuses the world with sustaining love. Ironically, in the light of Christianity's misunderstood history, it is, then, a religion of *carnal subscendence*, of God giving Godself in the many life webs of existence, not *otherworldly transcendence*, where God is pictured as cosmically remote from planetary well-being.

Today, many of the Swarthmore College students I teach now comprehend, bitterly, that our society's continued dependence on fossil-fuel burning to power our economy has driven us into a climatological cul-de-sac. These students, and their compatriots elsewhere, are leading the political movement in higher learning institutions to divest university endowments from the coal, oil, and gas extraction industry through nonviolent, direct-action takeovers and sit-ins of campuses everywhere. At Swarthmore College, I and others have encouraged our students to act out of the fullness of their liberal education—an education that stresses intellectual rigor and moral accountability. We believe it is wrong that Swarthmore's $2 billion endowment is significantly invested in the very industry that is destroying the living systems that make possible and viable human civilization as we know it. If Swarthmore's goal is to teach students to think critically and responsibly in a changing world, then it should not be surprising that these very same students now decry the college's investments in an industry that poses a mortal threat to planetary well-being.

Swarthmore students and I understand the importance of this historical moment. We will continue to use all of the tactical tools at our disposal to persuade Swarthmore's oversight board and administration to align all of its actions—including its financial dealings—to the one goal of moving the college and the wider society toward a healthy, verdant, fossil-fuel-free future. For my part, this is a painful time for me at my institution. My advocacy has frayed long-term friendships with administrative staff and some faculty. At times, I feel like a pariah among my professional colleagues. But now that a growing list of institutions, including the Rockefeller Brothers Fund, have pledged to rid themselves of their fossil-fuel holdings, we recognize how sad and ironic it is that Swarthmore College—a Quaker-heritage institution once lauded for its pathbreaking moral leadership in the abolitionist, suffragist, civil rights, and LGBTQ movements—has become hopelessly mired in the morally dubious and financially risky fossil-fuel business. Instead of leading the charge into a green future, hidebound Swarthmore is fighting a rearguard action to stop its inevitable transition into a postoil economy. We all know that this changeover is coming: either my college and the wider society will initiate this transition willingly, or nature will drive us to our knees and compel us to make this transition against our wills. So students and I have implored the college to align its assets with its values, to stand on the right side of history, and to use its enormous wealth for the common good by combating the scourge of human-caused climate change in our time.

Joining the campus divestment movement is rooted in my Christian animist experience of the world as sacred gift. This experience is an exercise in joy, if not always happiness; in hope, if not always optimism; in faith, if not always certitude; in vision, even when clear sight is not possible; in song, even when the music cannot be heard; and in moving forward, even when the way is not clear and the obstacles seem overwhelming. My animist yearning is that through the eyes of faith, the world will become enlivened again as chock-full of living souls—tree people, rock people, river people, bird people—all of whom urge us to comport ourselves toward them with dignity and compassion. In *Paradise Lost*, John Milton writes that "millions of spiritual creatures walk the earth unseen, both when we wake and when we sleep."[45] In *Aurora Leigh*, Elizabeth Barrett Browning imagines that "Earth's crammed with heaven, and every common bush afire with God."[46] And in *God's Grandeur*, Gerard Manley Hopkins sings that the "Holy Spirit over the bent World broods with warm breast and with ah! bright wings."[47] Are the poets right? Could this be so? Could Earth again be encountered as a living being full of spiritual creatures—the birds of the air, the fish of the sea, the lilies of the field, the rivers in the valleys, the rocks of the mountains, the trees of the forest, the sun and stars of the sky—and thereby deserving of our reverence and worthy of our protection? Could Earth once more be envisioned as crammed with heaven, and every common thing as afire with God? In an incantatory gesture, could the teeming wonder of the many life-forms among us be a reminder of the Spirit in the Earth—indeed, a God of beak and feathers?

If this could be so, then Earth would no longer be the deicidal boneyard for the winged God of Christian witness. If this could be so, then into the living waters, vital food chains, fertile soil, and the quickening fiery atmosphere, God, always threatened and vulnerable, would be seen as streaming forth to ensure life and vitality for all beings. If this could be so, then we could learn to care for the world with heartfelt abandon—even as the Gospels' bird-God would be viewed as pouring out Godself into the wonder and plenitude of our planetary commons. "When Jesus was baptized, the Holy Spirit descended upon him in bodily form as a dove" (Luke 3:21–22). If this could be so, then all things would be viewed as bearers of the kenotic sacred; each and every fleshy creature and natural element would be seen as a portrait of the avian God; and everything that is would be cherished as holy and blessed and good.[48]

# ACKNOWLEDGMENTS

I am full of gratitude to my friends, colleagues, and family who have helped me to bring to fruition my Christian animism project. Paul Ricoeur, one of my teachers at the University of Chicago Divinity School, writes that the journey to selfhood consists of "aiming at the 'good life' with and for others, in just institutions." I have been blessed to be affiliated with three such institutions that have encouraged my pursuit of the good life with and for others and, in that vein, have supported the publication of this book.

Swarthmore College has nurtured my personal and professional development since soon after I began my teaching career. Swarthmore prizes the liberal arts as essential to civic society and provides extraordinary material support for faculty to pursue their research interests. The college offers teaching scholars one- or two-term sabbaticals every four years—my sabbatical two years ago enabled me to jump-start this project—and it provides subventions from the Constance Hungerford Faculty Support Fund, which allowed me to pay for the moving illustrations in this book by James Larson and the superb index by Claire Splan.

As well, I have been fortunate to be a member of the Constructive Theology Workgroup, a long-standing community of liberation-oriented Christian thinkers whose regular colloquia—in which we read and critique each other's emerging work—have been the seedbed for this project. I am grateful to the erstwhile conveners and members of the workgroup—in particular, Wendy Farley, Mary McClintock Fulkerson, Marion Grau, Serene Jones, Catherine Keller, Paul Lakeland, Jim Perkinson, Shelly Rambo, Darby Ray, Stephen Ray, Laurel Schneider, Don Schweitzer, and Deanna Thompson—who created a unique habitat for nurturing our collective, and individual, theological visions in the workgroup.

The Forum on Religion and Ecology, spearheaded by Mary Evelyn Tucker and John Grim at Yale University, has been a third institutional collective essential to providing the conceptual and conference resources necessary for completing this research. In my Christian vocabulary, I consider Mary Evelyn and John to be bearers of the gift of apostleship—spiritual leaders, committed to the welfare of others, who live according to a sense of calling or mission. As friends and determined caretakers of the forum's network of conferences, publications, and information sharing, they have built together the now-emergent academic field of religion and ecology (along with Bron Taylor's inspired leadership of the cognate field of religion and nature) and firmly established it within the academy and the wider public sphere.

Swarthmore colleagues inside and outside my home department—Tariq al-Jamil, Yvonne Chireau, Giovanna DiChiro, Richard Eldridge, Steven Hopkins, Gwynn Kessler, Grace Ledbetter, Milton Machuca-Galvez, Helen Plotkin, Ellen Ross, Christy Schuetze, Val Smith, Jamie Thomas, José Vergara, Hansjacob Werlen, and Craig Williamson—were engaged and thoughtful dialogue partners in my efforts to complete this book. Steven and Adrienne Caddell-Hopkins were especially helpful in conversations about the book, along with three other couples with whom Audrey and I share meals and laughter: Todd and Aimee Drumm, Jon and Lisa Pahl, and Ken and Mary Gergen.

Tom Lay, my editor at Fordham University Press, carefully and tirelessly shepherded this project to completion. Eric Newman, managing editor, and Andrew Katz, copyeditor, devoted their talents and industry to improving my prose and general presentation. The academic editors of the Groundworks series—Brian Treanor and Forrest Clingerman—were early adopters of the book's potential and helped to shape my overall thesis and approach. Two anonymous readers for the press provided rich insights into the book's promise, as well as its many deficiencies, and I owe both of these readers my deepest thanks. On early-morning walks, David Eberly was incredibly generous in teaching me about the rich birdlife in our own Crum Woods watershed. My Swarthmore student assistant, Ian Palmer, offered a thoughtful evaluation of the book's message. And Matthew Eaton provided timely and detailed suggestions and critique regarding the book's individual chapters and overall scope.

Opportunities to give version of the book's chapters in the following learning environments were greatly appreciated: the Dean's Seminar in the Craft of Teaching at the University of Chicago Divinity School; the Center for the Study of World Religions at Harvard University; the Institute for Religion and Science at Chestnut Hill College; the Liberal Theologies Group at the American Academy of Religion; the Collegiate Peaks Forums Series in Salida, Colorado; the Winter Refresher Conference at

St. Andrew's College in Saskatoon, Canada; the Post-Traumatic Theology Conference at the Boston University School of Theology; the Living Cosmology Conference at Yale Divinity School; the Colloquium on Violence and Religion at the University of Northern Iowa; and the Ritual and Loving the Earth Conference at Ghost Ranch in Abiquiu, New Mexico.

My sister, Darice Wallace, has been a source of joy and stability throughout my life; her love of the natural world is reflected throughout this writing. My children, Katie and Chris Ross-Wallace, were bundles of grace in my life as children and now, as young adults, thoughtful pilgrims in their own journeys to selfhood. My final heartfelt word of gratitude goes to my wife, Audrey, to whom this book is dedicated and with whom I have joyously labored in the animist vineyard for many years. The still point in my turning world, Audrey daily and patiently enfleshes the numinous presence of the sacred within my own broken existence.

# NOTES

## INTRODUCTION: CROSSING THE SPECIES DIVIDE

1. Sean McGrath, "The Question Concerning Nature," in *Interpreting Nature: The Emerging Field of Environmental Hermeneutics*, ed. Forrest Clingerman, Brian Treanor, Martin Drenthen, and David Utsler (New York: Fordham University Press, 2014), 216.

2. Recent work on the Spirit in theology and philosophy that has informed my own thinking includes Sharon V. Betcher, *Spirit and the Obligation of Social Flesh: A Secular Theology for the Global City* (New York: Fordham University Press, 2013); Ada María Isasi-Díaz, Barbara A. Holmes, Serene Jones, Catherine Keller, Walter J. Lowe, Jim Perkinson, Mark I. Wallace, and Sharon D. Welch, "Spirit," in *Constructive Theology: A Contemporary Approach to Classical Themes*, ed. Serene Jones and Paul Lakeland (Minneapolis: Fortress, 2005), 239–78; Jacques Derrida, *Of Spirit: Heidegger and the Question*, trans. Geoffrey Bennington and Rachel Bowlby (Chicago: University of Chicago Press, 1989); and Adolf Holl, *The Left Hand of God: A Biography of the Holy Spirit*, trans. John Cullen (New York: Doubleday, 1998).

3. The term belongs to Mircea Eliade in *Shamanism: Archaic Techniques of Ecstasy*, trans. Willard R. Trask (New York: Pantheon, 1964), 156–58. Also see Jacques Derrida's neologism "divinanimality" for analyzing the animal-other as another name for God in *The Animal That Therefore I Am*, ed. Marie-Louise Mallet, trans. David Wills (New York: Fordham University Press, 2008), 132.

4. Matthew Hall, *Plants as Persons: A Philosophical Botany* (Albany: SUNY Press, 2011), 55–71. For further meditations on the emotional lives, communication strategies, and underground social networks of trees, see Peter Wohlleben, *The Hidden Lives of Trees: What They Feel, How They Communicate* (Vancouver: Greystone, 2016).

5. In this regard, see Lynn White Jr., "The Historical Roots of Our Ecologic Crisis," *Science* 155 (1967): 1203–7; Daniel Quinn, "Animism—Humanity's Original Religious Worldview," in *The Encyclopedia of Religion and Nature*, ed. Bron R. Taylor, 2 vols. (New York: Continuum, 2005), 1:81–83; Ed McGaa, *Mother Earth Spirituality: Native American Paths to*

*Healing Ourselves and Our World* (New York: Harper and Row, 1990); Vine Deloria Jr., "Sacred Places and Moral Responsibility," in *Worldviews, Religion, and the Environment: A Global Anthology*, ed. Richard C. Foltz (New York: Wadsworth, 2002), 83–91; and John A. Grim, "Indigenous Traditions and Deep Ecology," in *Deep Ecology and World Religions: New Essays on Sacred Ground*, ed. David Landis Barnhill and Roger S. Gottlieb (Albany: SUNY Press, 2001), 35–57.

6. This phrase has many meanings, but I intend it here as a gloss on Martin Luther's "Babylonian Captivity of the Church" from 1520, a polemical treatise intended to recover ancient biblical beliefs and practices over and against some of the excesses of the exclusionary sacramental polity of Western Christianity in the late medieval period. See Martin Luther, "The Babylonian Captivity of the Church," in *Three Treatises*, ed. Helmut T. Lehmann, trans. A. T. W. Steinhäuser, 2d ed. (Minneapolis: Fortress, 1990), 113–260.

7. Shawn Sanford Beck, *Christian Animism* (Alresford, U.K.: Christian Alternative Books, 2015), 8.

8. In this book, I sometimes capitalize "Earth" when referring to the entire planet, in the sense of the wide expanse of the natural world, while at other times I prefer the term "earth" in the sense of the "dirt" or "soil" or "matter" that makes up the surface of the planet.

9. See Aaron S. Gross, "The Question of the Creature: Animals, Theology and Levinas' Dog," in *Creaturely Theology: On God, Humans and Other Animals*, ed. Celia Deane-Drummond and David Clough (London: SCM, 2009), 121–37; Nathan M. Bell, "Environmental Hermeneutics with and for Others: Ricoeur's Ethics and the Ecological Self," in Clingerman et al., *Interpreting Nature*, 141–59; Sallie McFague, "Reflecting God," in *After God: Richard Kearney and the Religious Turn in Continental Philosophy*, ed. John Panteleimon Manoussakis (New York: Fordham University Press, 2006), 362–64; and Louise Westling, *The Logos of the Living World: Merleau-Ponty, Animals, and Language* (New York: Fordham University Press, 2014).

10. The postcolonial recovery of animism accords with recent work in *posthumanism* (see Carey Wolfe, *What Is Posthumanism?* [Minneapolis: University of Minnesota Press, 2010]; and Jennifer L. Koosed, ed., *The Bible and Posthumanism* [Atlanta: Society of Biblical Literature, 2010]) and the *new materialism* (see Jane Bennett, *Vibrant Matter: A Political Ecology of Things* [Durham, N.C.: Duke University Press, 2010]; and Clayton Crockett and Jeffrey W. Robbins, *Religion, Politics, and the Earth: The New Materialism* [New York: Palgrave Macmillan, 2012]). Related to postcolonial animism are new studies in animacy theory and animality in, for example, Mel Y. Chen, *Animacies: Biopolitics, Racial Mattering, and Queer Affect* (Durham, N.C.: Duke University Press, 2012).

11. Martin Buber, *I and Thou*, trans. Walter Kaufmann (New York: Charles Scribner's Sons, 1970), 57, 145.

12. The quote is from Levinas in Tamara Wright, Peter Hughes, and Alison Ainley, "The Paradox of Morality: An Interview with Emmanuel Levinas," in *The Provocation of Levinas*, ed. Robert Bernasconi and David Wood (London: Routledge, 1988), 169.

13. Jacques Derrida, *The Animal That Therefore I Am*, trans. David Wills (New York: Fordham University Press, 2008), 18.

14. For two comparative religious and theological works that analyze Christianity, among other religions, and sacred animals, in particular, see Paul Waldau and Kimberley Patton, eds., *A Communion of Subjects: Animals in Religion, Science, and Ethics* (New York: Columbia University Press, 2006); and Stephen D. Moore, ed., *Divinanimality: Animal Theory, Creaturely Theology* (New York: Fordham University Press, 2014).

15. John Grim, "Knowing and Being Known by Animals: Indigenous Perspectives on Personhood," in Waldau and Patton, *Communion of Subjects*, 379.

16. E. B. Tylor, *Primitive Culture: Researches into the Development of Mythology, Philosophy, Religion, Art, and Custom*, 2 vols. (1871; repr., New York: Gordon, 1974) 2:170–71.

17. Ibid., 1:385.

18. David L. Haberman, *People Trees: Worship of Trees in Northern India* (Oxford: Oxford University Press, 2013), 9.

19. Ibid., 20.

20. Graham Harvey, "Animism—A Contemporary Perspective," in Taylor, *Encyclopedia of Religion and Nature*, 1:81.

21. David Abram, *Becoming Animal: An Earthly Cosmology* (New York: Vintage, 2010), 47.

22. See "new animism" studies of human-nature intersubjectivity in Patrick Curry, "Grizzly Man and the Spiritual Life," *Journal for the Study of Religion, Nature and Culture* 4 (2010): 206–19; Priscilla Stuckey, "Being Known by a Birch Tree: Animist Refigurations of Western Epistemology," *Journal for the Study of Religion, Nature and Culture* 4 (2010): 182–205; and Abram, *Becoming Animal*.

23. George "Tink" Tinker, "The Stones Shall Cry Out: Consciousness, Rocks, Indians," *Wicazo Sa Review* 19 (2004): 105–25.

24. Graham Harvey, *Animism: Respecting the Living World* (New York: Columbia University Press, 1976), xi.

25. Harvey, "Animism—A Contemporary Perspective," 1:81 (emphasis added).

26. Bron Taylor, *Dark Green Religion: Nature Spirituality and the Planetary Future* (Berkeley: University of California Press, 2010), 178. In a footnote to

this quote, Taylor notes my earlier work in animist Christianity in *Finding God in the Singing River: Christianity, Spirit, Nature* (Minneapolis: Fortress, 2005) as "an exception that proves the rule" that Christianity and animism are categorically distinct. But if there are exceptions to a rule, then the rule itself should be questioned. In my judgment, and as I hope to demonstrate in this book, the age-old opposition between Christianity and animism trades on a false choice.

27. See "Geologian," Thomas Berry's website, accessed November 9, 2017, http://thomasberry.org/life-and-thought/about-thomas-berry/geologian.

28. Wallace, *Finding God in the Singing River*, 18.

29. Mark I. Wallace, *Fragments of the Spirit: Nature, Violence, and the Renewal of Creation* (New York: Continuum, 1996), 144.

30. Along with my *Finding God in the Singing River*, the category of Christian animism has resonated in a book by Shawn Sanford Beck (*Christian Animism*), blogs by Noel Moules (*Christian Animism*: http://www.christiananimism.com) and Stuart Masters ("Some Themes in Contemporary Christian Eco-Theology and Bible Scholarship," *A Quaker Stew*, January 13, 2017, http://aquakerstew.blogspot.com/2017/01/some-themes-in-contemporary-christian.html), and Jay Beck's ceremonial theater (Wild Goose Festival's Carnival de Resistance: http://wildgoosefestival.org/carnival-de-resistance/).

31. For example, see Marcella Althaus-Reid, *The Queer God* (London: Routledge, 2003); and Deryn Guest, Robert E. Goss, Mona West, and Thomas Bohache, eds., *The Queer Bible Commentary* (London: SCM, 2006).

32. The revival of the term *animism* in religious studies owes much to Nurit Bird-David's "'Animism' Revisited: Personhood, Environment, and Relational Epistemology," *Current Anthropology* 40 (1999): 67–91; and Graham Harvey's *Animism: Respecting the Living World*. See the insightful summary of this trend in Darryl Wilkinson, "Is There Such a Thing as Animism?," *Journal of the American Academy of Religion* 84 (2016): 1–23.

33. George Cajete, "Philosophy of Native Science," in *American Indian Thought: Philosophical Essays*, ed. Anne Waters (Oxford, U.K.: Blackwell, 2004), 50.

34. Wilkinson, "Is There Such a Thing as Animism?," 7.

35. Linda Hogan, "We Call It *Tradition*," in *The Handbook of Contemporary Animism*, ed. Graham Harvey (London: Routledge, 2014), 21.

36. Laura Hobgood-Oster, *Holy Dogs and Asses: Animals in the Christian Tradition* (Urbana: University of Illinois Press, 2008) 47.

37. Barbara Allen, *Animals in Religion: Devotion, Symbol, and Ritual* (London: Reaktion Books, 2016), 186.

38. Ibid., 235.

39. I am borrowing this phrase from Brian Massumi, *What Animals Teach Us about Politics* (Durham, N.C.: Duke University Press, 2014), 52.

40. In this regard, I am inspired by the work on "deep incarnation" in Niels Henrik Gregersen, ed., *Incarnation: On the Scope and Depth of Christology* (Minneapolis: Fortress Press, 2015); and Denis Edwards, *Ecology at the Heart of Faith: The Change of Heart That Leads to a New Way of Living on Earth* (Maryknoll, N.Y.: Orbis, 2006). Gregersen et al.'s "full-scope," rather than "strict-sense," view of incarnation—that God becoming one with Jesus is an "icon" of God's full immersion in creaturely existence—is extended with special reference to animal incarnation (and the risks to God in such incarnation) by Matthew Eaton's monistic, panincarnationist model of God and the world in "Enfleshing Cosmos and Earth: An Ecological Theology of Divine Incarnation," (Ph.D. diss., University of St. Michael's College of the Toronto School of Theology, 2016). On the incarnation as libertine or promiscuous, see Althaus-Reid, *Queer God*, 46–76; and Laurel C. Schneider, *Beyond Monotheism: A Theology of Multiplicity* (London: Routledge, 2008), 198–207.

41. David L. Clough, *On Animals*, vol. 1 of *Systematic Theology* (London: T. and T. Clark, 2012), 103.

42. Ibid., 102.

43. *Christian animism* is not *pantheism*, but it is also not homologous with its near-cognate term *panentheism*. Like animism, panentheism does speak to the abiding interrelationship of God and the world, but the idea that all-things-are-in-God does not necessarily entail the animistic notion, as I have defined it here apropos Harvey et al., that all things are alive, that all things are persons, and that all things are sacred. Likewise, while some forms of pantheism move away from the understanding of God as personal being, Christian animism retains this formulation and expands it to include the more-than-human personhood of God as well. Unlike pantheism, panentheism does carry the double valence of God's alternate dependence on and independence from the world. But unlike animism, it does not necessarily entail the ascription of living, sacred personhood to all beings—a critical biblical value that the category of Christian animism brings to the forefront. In spite of these caveats, in this book I use the term *panentheism*, but not *pantheism*, as closely akin but not equivalent to *Christian animism*.

44. The phrase belongs to Althaus-Reid in *Queer Theology*, 87.

45. Walter Lowe, *Theology and Difference: The Wound of Reason* (Bloomington: Indiana University Press, 1993), 11–13.

46. Sallie McFague, *The Body of God: An Ecological Theology* (Minneapolis: Fortress, 1993), 150.

47. Catherine Keller, *Cloud of the Impossible: Negative Theology and Planetary Entanglement* (New York: Columbia University Press, 2015), 7.

48. Buber, *I and Thou*, 130.

49. Brian Treanor, "Narrative and Nature: Appreciating and Understanding the Nonhuman World," in Clingerman et al., *Interpreting Nature*, 181–200.

### 1. SONG OF THE WOOD THRUSH

1. The Crum Woods of Swarthmore College is a large fragment of a once-great forest that extended across the eastern half of North America. Its rich habitat features a gentle stream called the Crum Creek (*crum* in Swedish means "crooked") that winds its way through steep slopes and dramatic outcrops flecked with mica and quartz. Today the woods consist of mixed hardwood trees, varieties of grasses and wildflowers, white-tailed deer and American eels, and many of the noteworthy birds featured in this book, including the wood thrush, pileated woodpecker, great blue heron, and common pigeon—the common pigeon being the particular bird most closely identified today with the dove highlighted in the Gospels' stories of Jesus' baptism. On the history and current state of the Crum Woods ecosystem, see R. E. Latham, D. B. Steckel, H. M. Harper, and D. C. Rosencrance, *Conservation and Stewardship Plan for the Crum Woods of Swarthmore College*, report for Swarthmore College by Natural Lands Trust, Media, and Continental Conservation, Rose Valley, PA, December 2003, https://www.swarthmore.edu/sites/default/files/assets/documents/crum-woods-stewardship-committee/Conservation_and_Stewardship_Plan_2003.pdf.

2. See an early study, using audiospectography, of overlapping, simultaneous notes singing by wood thrushes in Ohio in Donald J. Borror and Carl R. Reese, "Vocal Gymnastics in Wood Thrush Songs," *Ohio Journal of Science* 56 (1956): 177–82.

3. Henry David Thoreau, *The Heart of Thoreau's Journals*, ed. Odell Shepard (New York: Dover, 1961), 92.

4. Today's "forest church" movement resituates religious worship deep in the heart of wooded environments set apart from traditional built sanctuaries. The conventional ecclesiastical lexicon is going green: the semantic range of terms such as *sacred space* or *sanctuary* now extends beyond the category of "house of worship." These efforts to "rewild" Christianity are provocatively narrated in Fred Bahnson, "The Priest in the Trees: Feral Faith in the Age of Climate Change," *Harper's Magazine*, December 2016, 46–54. Also see Ric Hudgen's related blog, *Wild Church Network*, at https://www.wildchurchnetwork.com.

5. E. O. James, *The Worship of the Sky-God: A Comparative Study in Semitic and Indo-European Religion* (London: Athlone, 1963), 168.

6. Norman Wirzba's nature-based biblical theology is a forceful critique of humankind's degradation of nature and its concomitant displacement of God into the realm of the invisible. He writes, "The extent of our own mastery, and of our deformation of God, can be seen in the eclipse of nature. It is increasingly difficult to look at our environment and not see everywhere a reflection of ourselves and our own activity. . . . Though we may try, it is hard to discern where the hand of God ends and where the hand of humanity begins, which forces the question: Is our world the creation of God at all?" Norman Wirzba, *The Paradise of* God (Oxford: Oxford University Press, 2003), 91. God, as the distant sky-God, has rendered the God of creation no longer visible because we have remade the natural world in our own image. The eclipse of nature and the eclipse of God are the same thing. For Wirzba, just insofar as we understand Earth today as a commodified *object* under our manipulation and control, it is now almost impossible for us to see it as the gift of *creation* in the biblical sense.

7. In a paper on Genesis 1, Norman C. Habel comments on this parallel raptor-Spirit imagery in Deuteronomy: "The use of this verb [*merahefet*] in Deut. 32.11 speaks of an eagle 'hovering' over its young and spreading its wings—not in some fierce act of disturbance, but apparently to lift them up in the nurturing act of teaching them to fly." Norman C. Habel, "Geophany: The Earth Story in Genesis 1," in *The Earth Story in Genesis*, vol. 2 of *The Earth Bible*, ed. Norman C. Habel and Shirley Wurst (Sheffield, U.K.: Sheffield Academic, 2000), 37.

8. See Luise Schottroff, "The Creation Narrative: Genesis 1.1–2.5a," in *A Feminist Companion to Genesis*, ed. Athalya Brenner (Sheffield, U.K.: Sheffield Academic, 1993), 24–38.

9. Perhaps it is easier to rethink existence as sacred in human and animal terms, but the woody plant world should be equally compelling, as the story of the burning bush indicates. Vegetative, arboreal forms of divinity characterize all of the world's religions, including Christianity. The Buddha's enlightenment took place under a fig tree, the time-honored Bodhi Tree. Sacred banyan and peepal trees are living beings worthy of worship and compassion in India. The Lakota sun dance ceremony revolves around a cottonwood tree as a symbol of the sacred's abiding presence in all things. And Jesus' saving death occurred on a tree, species unknown, as testified to by Peter: "The God of our fathers raised up Jesus, whom you put to death by hanging him from a tree" (Acts 5:30). For a moving meditation on the religious and environmental significance of the crucifixion of the particular tree on which Jesus died at Golgotha, see Stephanie Kaza, "House of

Wood," in *This Sacred Earth: Religion, Nature, Environment*, ed. Roger S. Gottlieb (New York: Routledge, 1995), 41–43.

10. Edward W. Desmond, "Interview with Mother Teresa," *Time*, December 4, 1989.

11. Karl Barth, *Church Dogmatics*, vol. 1, pt. 1, *The Doctrine of the Word of God, Part 1*, trans. G. W. Bromiley (1936; repr., Edinburgh: T. & T. Clark, 1975), 55.

12. See the full text in "The Gospel of the Ebionites," in *Lost Scriptures: Books That Did Not Make It into the New Testament*, by Bart D. Ehrman (Oxford: Oxford University Press, 2003), 12–14.

13. See the history of these images in Fred S. Kleiner and Christin J. Mamiya, eds., *Gardner's Art through the Ages: The Western Perspective*, 3 vols. (Belmont, Calif.: Thomson/Wadsworth, 2005).

14. To this point, see W. Stewart McCullough's entries on *peristera*, which he alternately translates as "dove" and "pigeon," in his articles by these same names, in *The Interpreter's Dictionary of the Bible*, ed. George Arthur Buttrick, 5 vols. (Nashville, Tenn.: Abingdon, 1962), 1:866–67, 3:810.

15. Debbie Blue, *Consider the Birds: A Provocative Guide to Birds of the Bible* (Nashville, Tenn.: Abingdon, 2013), 9–10.

16. See Mircea Eliade, *Shamanism: Archaic Techniques of Ecstasy*, trans. Willard R. Trask (New York: Pantheon, 1964), 156–58.

17. Tikva Frymer-Kensky, *In the Wake of the Goddesses: Women, Culture, and the Biblical Transformation of Pagan Myth* (New York: Fawcett Columbine, 1993), 154.

18. Michael York, "Shamanism—Traditional," in *The Encyclopedia of Religion and Nature*, ed. Bron R. Taylor, 2 vols. (New York: Continuum, 2005), 2:1534.

19. Howard Eilberg-Schwartz, *The Savage in Judaism* (Bloomington: Indiana University Press, 1990), 236.

20. Ibid.

21. Eliade, *Shamanism*, 157 and following. Eliade details how important animal spirits are in historical shamanistic practices worldwide.

22. Frymer-Kensky, *In the Wake of the Goddesses*, 153.

23. Ibid., 153–54.

24. In addition to Jesus' pronouncement that he is the new Mosaic serpent, three other New Testament texts underscore the revival of snake religion in Jesus' time. In Matthew, Jesus valorizes the serpent as a symbol of sharp-witted intelligence, comparing the shrewdness of snakes and the innocence of doves as personality characteristics necessary for discipleship (10:16). In other texts, snake handling, as it is called today, was a sign of

divine power among Jesus' followers. In the longer ending to Mark, true believers will not be injured when "they pick up snakes in their hands" (16:18), and in Acts, Paul handles a snake without injury over a burning fire, signaling to his astonished onlookers "that he was a god" (28:3–6). My thanks to Matthew Eaton for noting these parallel snake passages in the Christian scriptures.

25. Clara Sue Kidwell, Homer Noley, George E. "Tink" Tinker, *A Native American Theology* (Maryknoll, N.Y.: Orbis Books, 2001), 81. This passage comes from the chapter on Christology with the note, "George E. 'Tink' Tinker took primary responsibility for this chapter" (186).

26. Lynn White Jr., "The Historical Roots of our Ecologic Crisis," *Science* 155 (1967): 1205.

27. Pat Zukeran, *World Religions through a Christian Worldview* (Richardson, Tex.: Probe Ministries, 2008), 43.

28. Robert A. Sirico, "The New Spirituality," *New York Times Magazine*, November 23, 1997.

29. Belden C. Lane, *Ravished by Beauty: The Surprising Legacy of Reformed Spirituality* (Oxford: Oxford University Press, 2011), 125.

30. Ibid.

31. To note two examples in this regard, see Nicole A. Roskos, "Felling Sacred Groves: Appropriation of a Christian Tradition for Antienvironmentalism," in *Ecospirit: Religions and Philosophies for the Earth*, ed. Laurel Kearns and Catherine Keller (New York: Fordham University Press, 2007), 483–92, on Christian groups dedicated to land and forest conservation, for example, Christians for the Mountains, the Religious Campaign for Forest Conservation, and the Natural Religious Partnership for the Environment; and regarding historical attempts by Quakers to save tens of thousands of acres of first- and second-growth rain forest in Costa Rica, see Nalini M. Nadkarni and Nathaniel T. Wheelwright, *Monteverde: Ecology and Conservation of a Tropical Cloud Forest* (Oxford: Oxford University Press, 2000).

32. The term *panentheism*, as noted in Chapter 1, is a variation on *pantheism* and has been advanced by numerous thinkers, from modern philosophers G. W. F. Hegel and Alfred North Whitehead to contemporary theologians Paul Tillich, Jürgen Moltmann, Catherine Keller, and Sallie McFague. McFague defines *panentheism* in this way: "Everything that is is *in* God and God is *in* all things and yet God is not identical with the universe." McFague, *The Body of God: An Ecological Theology* (Minneapolis: Fortress, 1993), 149. Paul's vision of God's all-encircling love is resonant with the panentheistic, animistic vision of God as the self-giving ground of all beings evoked by the Greek thinkers he cites, Epimenides and Aratus.

33. "The Martyrdom of Polycarp," in *The New Testament and Other Early Christian Writings*, ed. Bart D. Ehrman (Oxford: Oxford University Press, 2004), 187 (emphasis added).

34. On the phenomenology of spirit possession, see Karen McCarthy Brown's analysis in relation to Haitian vodou in *Mama Lola: A Vodou Priestess in Brooklyn* (Berkeley: University of California Press, 1991), esp. 350–64.

35. Leslie E. Sponsel, *Spiritual Ecology: A Quiet Revolution* (Santa Barbara, Calif.: Praeger, 2012), 9.

36. Stephen Jay Gould, *Eight Little Piggies: Reflections in Natural History* (New York: Norton, 1993), 87.

37. The Rebbe of Chernobyl, quoted in Arthur Waskow, "And the Earth Is Filled with the Breath of God," *Cross Currents* 47 (1997): 355.

38. See Mark Schwartz, "Plant Diversity Threatened by Climate Change, Greenhouse Gas Buildup, Study Finds," *Stanford Report*, June 18, 2003, http://news.stanford.edu/news/2003/june18/jasperplants-618.html.

39. A telling example of this change is the rapid expansion of "ghost forests" in southern New Jersey near my home in the Delaware River watershed (Philadelphia). In recent years, tens of thousands of acres of Atlantic white cedar woodlands—including native bird, animal, and plant and wildflower habitats—have been lost due to rising seawater encroaching on coastal freshwater aquifers and swamps. See Frank Kummer, "'Ghost Forests' Show How Salt Water Is Advancing," *Philadelphia Inquirer*, May 22, 2107, A1, A9.

40. See Elisabeth Rosenthal and Andrew C. Revkin, "Science Panel Says Global Warming Is 'Unequivocal,'" *New York Times*, February 3, 2007 A1, A5. Also see James Gustave Speth, *Red Sky at Morning: America and the Crisis of the Global Environment* (New Haven, Conn.: Yale University Press, 2004); and Elizabeth Kolbert, *Field Notes from a Catastrophe: Man, Nature, and Climate Change* (New York: Bloomsbury, 2006).

41. On saving what we love, see Jack Turner, *The Abstract Wild* (Tucson: University of Arizona Press, 1996).

42. Wendell Berry, "The Peace of Wild Things," in *Earth Prayers from Around the World*, ed. Elizabeth Roberts and Elias Amidon (San Francisco: HarperSanFrancisco, 1991), 102.

### 2. THE DELAWARE RIVER BASIN

1. Tom Wilber, *Under the Surface: Fracking, Fortunes, and the Fate of the Marcellus Shale* (Ithaca, N.Y.: Cornell University Press, 2012), 2.

2. Carol French, "Dairy Farmer," in *Shalefield Stories: Personal Testimonies and Collected Stories* (Homestead, Pa.: Steel Valley Printers, 2013), 12. French's "Dairy Farmer" is the same narrative recounted to my tour group;

I am using the published version of her story as the basis of my group's conversation with her. For a video version of the same story, see urbandisasterrecords, "'From the Front Lines' Carol French, Bradford Co, PA," YouTube, January 3, 2013, http://www.youtube.com/watch?v=EBfeyBzsjDo.

3. French, "Dairy Farmer," 13.

4. Ibid., 12.

5. Wilber, *Under the Surface*, 119.

6. On former vice president Dick Cheney's attempts to promote a favorable, unregulated business climate for oil and gas production in the United States, see Steve Coll, *Private Empire: ExxonMobil and American Power* (New York: Penguin, 2012), 67–92.

7. EPA, "Assessment of the Potential Impacts of Hydraulic Fracturing for Oil and Gas on Drinking Water Resources," external review draft, June 2015, http://www.epa.gov/hfstudy.

8. French, "Dairy Farmer," 13.

9. Philadelphia Water Department, "Marcellus Shale Drilling in the Delaware River Basin," accessed November 9, 2017, http://www.phillywatersheds.org/marcellus-shale-drilling-delaware-river-basin.

10. The term "root metaphor" was coined by the philosopher Owen Barfield. Orienting metaphors are not mere figures of speech but expressions of the fundamental cognitive and emotional dispositions toward the world within particular social groups. Root metaphors are indices to how people make sense of and participate in the world around them. Root metaphors are founded in what Barfield calls human beings' "original participation" in the natural and social worlds they inhabit. See Barfield, *Saving the Appearances: A Study in Idolatry* (New York: Harcourt Brace Jovanovich, 1962), 116–25.

11. See this analysis in Martin Heidegger, "The Question Concerning Technology," in *The Question Concerning Technology and Other Essays*, trans. William Lovitt (New York: Harper Colophon, 1977), 3–35. While I find Heidegger's writings about an Earth-centered philosophy of technology and dwelling to be positive and instructive, the monstrous reality of his membership in the National Socialist Party at the height of World War II and the Final Solution, and his refusal to publicly critique in any meaningful way this decision, must always be accounted for in any critical recovery of Heidegger's thought such as mine. For a balanced assessment of Heidegger's intellectual project in the light of his Nazi fidelities, see Michael E. Zimmerman, *Heidegger's Confrontation with Modernity: Technology, Politics, and Art* (Bloomington: Indiana University Press, 1990), 222–47.

12. Heidegger, "Question Concerning Technology," 11.

13. Heidegger uses the German verb *stellen* to describe "setting-upon," which also has the connotations of ordering, arranging, challenging, and calling forth. The contrast he makes is between *hervorbringing* (bringing-forth; attentive participation in nature's own patterns and relationships) over and against *stellen* (making demands on nature to produce goods that serve human ends with little attention to organic patterns or in regard to the collateral damage that such demands make on the wider biological world).

14. Heidegger, "Question Concerning Technology," 6.

15. See this discussion at Accademia.org, "Michelangelo's Prisoners or Slaves," accessed November 9, 2017, http://www.accademia.org/explore-museum/artworks/michelangelos-prisoners-slaves/.

16. Heidegger, "Question Concerning Technology," 7.

17. Ibid., 16.

18. Martin Heidegger, "Building Dwelling Thinking," in *Basic Writings*, rev. and exp. ed., ed. David Farrell Krell (New York: HarperCollins, 1993), 355.

19. Ibid., 347–63.

20. Norman Wirzba, *The Paradise of God* (Oxford: Oxford University Press, 2003), 147. Also see Janine M. Benyus, *Biomimicry: Innovations Inspired by Nature* (New York: William Morrow, 1997).

21. The philosopher Richard Eldridge makes this point. He argues that animist and nonanimist worldviews depend on the cultural assumptions of one's social group. Eldridge does not argue for the truth of animism per se but rather that the materialist presumption against animism is culturally assumed, not rationally self-evident. He writes, "No practice-independent facts—if there are any such things—force either animist or materialist attitudes on us. Instead, which practices, together with their projections of theoretical entities and their ways of being, one inhabits is decisive for the attitudes one will have." Eldridge, "Is Animism Alive and Well?," in *Can Religion Be Explained Away?*, ed. D. Z. Phillips (New York: St. Martin's, 1996), 17. Moreover, he argues that by positing a worldview in which all beings are ensouled, animism offers modern persons an ennobling perspective on ritual, art, morality, and religion.

22. See the philosopher Paul Ricoeur's analysis of Kant's tripartite philosophy in "Hope and the Structure of Philosophical Systems," in *Figuring the Sacred: Religion, Narrative, and Imagination*, ed. Mark I. Wallace, trans. David Pellauer (Minneapolis: Fortress, 1995), 203–16.

23. Mary Douglas, *Purity and Danger: An Analysis of the Concepts of Pollution and Taboo* (New York: Routledge Ark, 1966), 121.

24. Ibid.

25. Ibid.

26. Julia Kristeva, *Powers of Horror: An Essay on Abjection*, trans. Leon S. Roudiez (New York: Columbia University Press, 1982), 69.

27. Maintaining ritual purity was an important aspect of religious and cultural life within first-century Israelite religion. It appears that the act of spitting—or spitting onto another person while being impure oneself—was deemed to be a polluting force. The Levitical purity code makes this point regarding ceremonial uncleanliness: "If the one with the [unclean] discharge spits on persons who are clean, then they shall wash their clothes, and bathe in water, and be unclean until the evening" (Leviticus 15:8). The meaning of this text regarding spitting is unclear. Does it mean that spittle is simply the medium by which a polluting contagion is communicated? Or does it mean that spitting per se is another form of a ritually unclean discharge that continues the impurity of the original unclean discharge itself? The text is not clear in this regard, but most likely spitting in the Hebrew Bible and the halakhic legal codes of Jesus' time was regarded as a source of ritual pollution because of its association with other forms of impure bodily discharges and fluids.

28. Kristeva, *Powers of Horror*, 3.

29. Cynthia D. Moe-Lobeda analyzes the telling and inextricable link between our despoilment of Earth and destruction of human communities as an exercise in "moral oblivion," in *Resisting Structural Evil: Love as Ecological-Economic Vocation* (Minneapolis: Fortress, 2013).

30. René Girard, *Things Hidden since the Foundation of the World*, trans. Stephen Bann and Michael Metteer (Stanford, Calif.: Stanford University Press, 1987), 157–58. Also see Richard A. Cohen, "Is René Girard's *Things Hidden since the Foundation of the World* a Gnostic Theology?" (unpublished paper, 2008); Cohen argues against Girard's supersessionist interpretation of Judaism as an incomplete religion that is then fulfilled by Christianity, which effectively supplants, to use a familiar phrase, the "Old Testament God of violence" with the "New Testament God of love."

31. For an overview of Girard's theory of mimetic desire and the scapegoat mechanism, see Mark I. Wallace and Theophus H. Smith, "Editors' Introduction," in *Curing Violence*, ed. Wallace and Smith (Sonoma, Calif.: Polebridge, 1994), xvii–xxvi; and Robert G. Hamerton-Kelly, "Religion and the Thought of René Girard: An Introduction," ibid., 3–24.

32. See René Girard, *Violence and the Sacred*, trans. Patrick Gregory (Baltimore: Johns Hopkins University Press, 1977), 39–67, 119–68; and Girard, *The Scapegoat*, trans. Yvonne Freccero (Baltimore: Johns Hopkins University Press, 1986), 44–75.

33. Girard, *Things Hidden since the Foundation of the World*, 220.

34. On the problem of Girard's androcentrism and consistent use of female examples for articulating his mimetic theory, see Nancy Jay, *Throughout Your Generations Forever: Sacrifice, Religion, and Paternity* (Chicago: University of Chicago Press, 1992). I am grateful to Audrey Wallace for bringing this point to my attention.

35. Girard, *Things Hidden since the Foundation of the World*, 220.

36. These two Girard quotes are from Rebecca Adams, "Violence, Difference, Sacrifice: A Conversation with René Girard," *Religion and Literature* 25 (1993): 9–33; Adams argues that Girard's discussion of positive mimesis is an underdeveloped aspect of his fundamental anthropology. Also see René Girard, *Battling to the End: Conversations with Benoît Chantre* (East Lansing: Michigan State University Press, 2009), 120–35. On the challenges of Girard's appeal to good mimesis in the light of his postmodern thesis that the subject's desires are never innate but intersubjectively constructed by others, see Martha J. Reineke, "After the Scapegoat: René Girard's Apocalyptic Vision and the Legacy of Mimetic Theory," *Philosophy Today* 55 (2011): 63–75; and Stephen L. Gardner, "The Deepening Impasse of Modernity," *Society* 47 (2010): 452–60 (review of Girard's *Battling to the End: Conversations with Benoît Chantre*).

37. The phrase belongs to Rabanus Maurus, a German Benedictine monk from the ninth century CE. It is known as a universal prayer of Christianity as a plea for the gift of the Spirit in the life of the community. Rabanus's poem "Come Holy Spirit" has been used in Gregorian chant, the Catholic catechism, and Gustav Mahler's Eighth Symphony.

38. On the biochemical cycles necessary for everyday respiration, see Rebecca Harman, *Carbon-Oxygen and Nitrogen Cycles: Respiration, Photosynthesis, and Decomposition (Earth's Processes)* (Portsmouth, NH: Heinemann, 2005).

39. In this regard, I have also been inspired by Richard Louv, *The Last Child in the Woods: Saving Our Children from Nature-Deficit Disorder* (Chapel Hill, N.C.: Algonquin Books, 2005).

40. See further discussion in my "Green Mimesis: Girard, Nature, and the Promise of Christian Animism," *Contagion: Journal of Violence, Mimesis, and Culture* 21 (2014): 1–14.

41. The phrase "ghost bird" belongs to Scott Weidensaul in his discussion of the historical depredation of, and contemporary search for, the ivorybill woodpecker (which now, sadly, appears to be extinct). See Weidensaul, *The Ghost with Trembling Wings: Science, Wishful Thinking, and the Search for Lost Species* (New York: North Point, 2002), 38–64.

42. Ibid., 49.

### 3. WORSHIPPING THE GREEN GOD

1. John Seed, Joanna Macy, and Pat Fleming, *Thinking like a Mountain: Towards a Council of All Beings* (Gabriola Island, B.C.: New Society, 1988); and see Julia Butterfly Hill's forest-based spiritual practice in her *The Legacy of Luna: The Story of a Tree, a Woman, and the Struggle to Save the Redwoods* (San Francisco: HarperSanFrancisco, 2000).

2. John B. Cobb Jr., "Protestant Theology and Deep Ecology," in *Deep Ecology and World Religions: New Essays on Sacred Ground*, ed. David Landis Barnhill and Roger S. Gottlieb (Albany: SUNY Press, 2001), 223.

3. Richard Bauckham, *Living with Other Creatures: Green Exegesis and Theology* (Waco, Tex.: Baylor University Press, 2011), 13.

4. Ibid., 52.

5. Paul Santmire, *The Travail of Nature: The Ambiguous Ecological Promise of Christian Theology* (Philadelphia: Fortress, 1985), 182. Also see Santmire, *Nature Reborn: The Ecological and Cosmic Promise of Christian Theology* (Minneapolis: Fortress, 2000), for his further analysis of the deep ambiguity in Christianity regarding the value and sacred purpose of the created order of things.

6. Albert Schweitzer, *The Quest of the Historical Jesus: A Critical Study of Its Progress from Reimarus to Wrede*, trans. W. Montgomery (London: Adam and Charles Black, 1910), 396–401.

7. Pieter F. Craffert, *The Life of a Galilean Shaman: Jesus of Nazareth in Anthropological-Historical Perspective* (Eugene, Ore.: Wipf and Stock, 2008), 28.

8. Stephen W. Lewis, *Landscape as Sacred Space: Metaphors for the Spiritual Journey* (Eugene, Ore.: Wipf and Stock, 2005), 91–99.

9. Augustine's unparalleled focus on the meaning of God's goodness in creation resounds repeatedly in his many publications. In addition to his short treaties, biblical expositions, and sermons, this focus is apparent throughout his three great treatises on Christian faith—*Confessions*, *On Christian Doctrine*, and *The City of God*—as well as his three multivolume works on Genesis: *A Commentary on Genesis: Two Books against the Manicheans*, *The Literal Meaning of Genesis: An Unfinished Book*, and *The Literal Meaning of Genesis: A Commentary in Twelve Books*.

10. Augustine, *Confessions*, trans. Garry Wills (New York: Penguin, 2002), bk. 7, chap. 18.

11. Augustine, *The City of God*, trans. Marcus Dods, Great Books of the Western World, ed. Robert Maynard Hutchins, vol. 18 (Chicago: Encyclopedia Britannica, 1952), 343 (bk. 12, chaps. 1–2).

12. Based on a comprehensive reading of Paul, my interpretation of his claim that humankind participates in Adam's lineage is *mimetic*, in the

Girardian sense of imitating a model, not *biological*, in the sense of a bad seed, so to speak, transmitted through sexual relations. But following Augustine, historical Christian thought made a wrong turn in this regard. Paul is not making a microcellular observation. Rather, his meaning is typological, not sexological: sin is socially, not embryonically, transmitted from one generation to the next in continuity with humankind's historical antihero, Adam. Peer influence is all determinative for Paul, not sexual intercourse. His point is that human beings' tendency to "sin" (*hamartia* in Greek simply means "to miss the mark" or "fall short," albeit sometimes terribly in the sense of a fatal flaw, so Greek tragedy) just means that we radically shape one another's sometimes wayward behaviors in our group interactions with one another. For Paul, the primary influence in this process of cultural (not biological) formation is the story of Adam, the negative exemplar of the Christian narrative. In Romans 5:14, Paul says that Adam is a "type [*tupos* in Greek means a figure or a model] of the one that was to come," namely, Jesus. Adam, then, is the archetype of human beings' predilection to engage in self-destructive behavior—a trait that is handed down to subsequent generations as they become adept at emulating the legendary woebegone father of the human race. Paul says nothing about sex in his exposition of Adam's eventual global transgression, only that Adam is the cultural model (Greek *tupos*) of all subsequent (sinful) human behavior.

13. Augustine, *City of God*, 366 (bk. 13, chap. 14).

14. Elaine Pagels, *Adam, Eve, and the Serpent: Sex and Politics in Early Christianity* (New York: Random House, 1988), 109.

15. Augustine, *City of God*, 366 (bk. 13, chap. 14).

16. Augustine, *Confessions*, bk. 2, chap. 1. My study of Augustine's disparagement of bodily passion is "Early Christian Contempt for the Flesh and the Woman Who Loved Too Much in the Gospel of Luke," in *The Embrace of Eros: Bodies, Desires and Sexuality in Christianity*, ed. Margaret D. Kamitsuka (Minneapolis: Fortress, 2010), 33–49.

17. Augustine, *Confessions*, bk. 4, chap. 13.

18. Ibid., bk. 4, chap. 14.

19. Augustine, *The Literal Meaning of Genesis*, trans. John Hammond Taylor, S.J., 2 vols. (New York: Newman, 1982) 1:41 (bk. 1, chap. 18) (emphasis added).

20. Augustine, *City of God*, 611 (bk. 12, chap. 24).

21. Hildegard of Bingen, *Scivias*, trans. Mother Columba Hart and Jane Bishop (New York: Paulist, 1979), 150 (2.1).

22. Ibid., 418 (3.7.9).

23. Ibid., 162 (.2.2).

24. Hildegard of Bingen, *Hildegard von Bingen's Physica: The Complete English Translation of Her Classic Work on Health and Healing*, trans. Priscilla Throop (Rochester, Vt.: Healing Arts, 1998), 192.

25. Here see the entry on "pelican" in *The Oxford Dictionary of the Christian Church*, 2d ed., ed. F. L. Cross and E. A. Livingstone (Oxford: Oxford University Press, 1983), 1059. On the pelican symbolism in the history of Christianity, see Debbie Blue's excellent chapter on the pelican in *Consider the Birds: A Provocative Guide to Birds of the Bible* (Nashville, Tenn.: Abingdon, 2013), 21–38.

26. In this regard, see John V. Taylor, *The Go-Between God: The Holy Spirit and the Christian Mission* (London: SCM Press, 1975).

27. Elizabeth Dryer, "An Advent of the Spirit: Medieval Mystics and Saints," in *Advents of the Spirit: An Introduction to the Current Study of Pneumatology*, ed. Bradford E. Hinze and D. Lyle Dabney (Marquette, Wisc.: Marquette University Press, 2001), 134.

28. Hildegard of Bingen, *The Book of the Rewards of Life (Liber Vitae Meritorum)*, trans. Bruce W. Hozeski (Oxford: Oxford University Press, 1994), 135.

29. Hildegard of Bingen, *Hildegard von Bingen's Physica*, 9.

30. Ibid., 102–3.

31. Ibid., 103.

32. See this history in Steven Palmer and Iván Molina, eds., *The Costa Rica Reader: History, Culture, Politics* (Durham, N.C.: Duke University Press, 2004).

33. For a full overview of this early period to the present, see Asociación de Amigos de Monteverde, *Monteverde Jubilee Family Album, April 19, 1951–April 19, 2001* (Monteverde, Puntarenas, Costa Rica: Asociación de Amigos de Monteverde, 2001). For an extended interview with Marvin Rockwell, one of the jailed young men and early Fairhope founders of the community, see Discoverycostarica, "Interview with Mr. Marvin Rockwell—Monteverde Costa Rica—Part 1," YouTube, July 10, 2013, http://www.youtube.com/watch?v=IVSBfUis7Po; and Discoverycostarica, "Interview with Mr. Marvin Rockwell—Monteverde Costa Rica—Part 2," YouTube, July 10, 2013, http://www.youtube.com/watch?v=euDeaEUCQUg.

34. See the discussion in Nalini M. Nadkarni and Nathaniel T. Wheelwright, *Monteverde: Ecology and Conservation of a Tropical Cloud Forest* (New York: Oxford University Press, 2000).

35. See more such comments by Wolf Guindon, also one of the previously incarcerated Quakers and community founders, in Kay Chornook and Wolf Guindon, *Walking with Wolf: Reflections on a Life Spent Protecting the Costa Rican Wilderness* (Hamilton, Ont.: Wandering Words, 2008).

36. Note the divide between the historical opposition of Quakers to consecrated buildings, or "steeple houses," as inimical to the immediacy of Christ's presence in individual worshippers and the nineteenth-century return of many American mid- and far-western Quaker congregations to more traditionally Protestant church buildings (aka "Friends Churches"), in the entry "Friends, Religious Society of" in Cross and Livingstone, *Oxford Dictionary of the Christian Church*, 538.

37. Shannon McIntyre, "A Quaker Meetinghouse in Costa Rica," *Timber Framing* 108 (June 2013): 22.

## 4. "COME SUCK SEQUOIA AND BE SAVED"

1. Muir's early autobiographical books *The Story of My Boyhood and Youth* and *My First Summer in the Sierra* can be found in *John Muir: Nature Writings*, ed. William Cronon (New York: Library of America, 1997), 1–146, 147–310.

2. John Muir, "The American Forests," from *Our National Parks, 1901*, ibid., 720.

3. John Muir, "Hetch Hetchy Valley," ibid., 814.

4. See Jedediah Purdy, "Environmentalism's Racist History," *New Yorker*, August 13, 2015, https://www.newyorker.com/news/news-desk/environmentalisms-racist-history; Mark David Spence, *Dispossessing the Wilderness: Indian Removal and the Making of the National Parks* (Oxford: Oxford University Press, 1999); and Paul Outka, *Race and Nature from Transcendentalism to the Harlem Renaissance* (New York: Palgrave Macmillan, 2008).

5. Spence, *Dispossessing the Wilderness*, 101–13.

6. Muir, *My First Summer in the Sierra*, 184.

7. See David S. Heidler and Jeanne T. Heidler, *Indian Removal: A Norton Casebook* (New York: Norton, 2007), 45–47.

8. Spence, *Dispossessing the Wilderness*, 103.

9. Muir, *My First Summer in the Sierra*, 285.

10. Purdy, "Environmentalism's Racist History."

11. Carolyn Merchant, "Shades of Darkness: Race and Environmental History," *Environmental History* 8 (July 2003): 381.

12. Muir's settler-colonialist blind spot regarding Indian removal in the establishment of the national parks system is deeply troubling. Today, the rationale for this forced dislocation of traditional peoples is still trotted out in histories of the parks movement. Indeed, these histories parse the U.S. government's parks-system land acquisitions as necessary "sacrifices" that first peoples (willingly?) made in order to ensure the "unique triumph" of the system. This "green washing" of the national park movement's origins

could not be further from the truth. As *National Geographic* puts it, "Sacrifices were made as the system grew to include today's 392 national parks, monuments, battlefields, seashores, recreation areas, and other areas. Many native peoples were displaced. . . . But in the final analysis, America's system of national parks became a unique triumph." See "U.S. National Parks—In the Beginning," *National Geographic*, May 26, 2010, http://travel.nationalgeographic.com/travel/national-parks/early-history/.

13. Mark Stoll, *Inherit the Holy Mountain: Religion and the Rise of American Environmentalism* (Oxford: Oxford University Press, 2015), 175.

14. Northrop Frye, *The Great Code: The Bible and Literature* (New York: Harcourt Brace Jovanovich, 1981), 40.

15. Muir, *Story of My Boyhood and Youth*, 20.

16. Jeanne Carr was one of Muir's closest friends, and it is the letters addressed to her that I will give special attention to in my following exposition of Muir's thought. Carr was assistant superintendent of the California public school system in the 1870s and 1880s and Muir's regular confidante and supporter. Along with her husband, Ezra, she nurtured Muir's literary and political labors on behalf of the wider Yosemite area. Many of Muir's most moving and intimate letters about God and nature were written to Jeanne Carr (aka Mrs. Ezra S. Carr), his mentor and literary muse.

17. Anne Rowthorn notes, in her beautiful collection of Muir's inspirational writings, the biblical rootedness of Muir's rhetoric, as well as his penchant for capitalizing key terms in order to signal the interchangeably of words such as *God*, *nature*, and *beauty*. She states, "Raised as a Christian, Muir never renounced his orthodox roots. Many of his writings have biblical overtones, and he even borrowed some scriptural phrases in his writings. . . . Muir often capitalized the words nature, beauty, love, soul, and universe, just as he capitalized the word god. For him, the perfect synonym for God was Beauty." Rowthorn, *The Wisdom of John Muir: 100+ Selections from the Letters, Journals, and Essays of the Great Naturalist* (Birmingham, Ala.: Wilderness, 2012), 34.

18. John Muir to Miss Katherine Merrill Graydon, February 5, 1880, in *John Muir: His Life and Letters and Other Writings*, ed. Terry Gifford (London: Bâton Wicks; Seattle: Mountaineers, 1996), 250.

19. John Muir to David Muir, March 20, 1870, in *Life and Letters of John Muir*, 1:209.

20. John Muir, *John of the Mountains: The Unpublished Journals of John Muir*, ed. Linnie Marsh Wolfe (Boston: Houghton Mifflin, 1938), 47.

21. See Forrest Clingerman, "Reading the Book of Nature: A Hermeneutical Account of Nature for Philosophical Theology," *Worldviews* 13 (2009): 72–91; Peter Harrison, *The Bible, Protestantism, and the Rise of Natural Science*

(New York: Cambridge University Press, 2001); David C. Lindberg and Ronald L. Numbers, eds., *God and Nature: Historical Essays on the Encounter between Christianity and Science* (Berkeley: University of California Press, 1986); and Olaf Pedersen, *The Book of Nature* (South Bend, Ind.: University of Notre Dame Press, 1992).

22. John Muir to Jeanne Carr, January 21, 1866, in *Life and Letters of John Muir*, 1:147.

23. John Muir to David Muir, April 10, 1870, ibid., 1:218.

24. John Muir to Jeanne Carr, April 3, 1871, ibid., 1:249.

25. John Muir to Jeanne Carr, February or March 1871, ibid., 1:242.

26. Mark 1:12–13, in *The Message: The Bible in Contemporary Language*, trans. Eugene H. Peterson (Carol Stream, Ill.: NavPress, 2002).

27. Muir, *John of the Mountains*, 82. One of the sad footnotes to Muir's encounters with the California grizzly, a subspecies of the wider family of grizzly bears, is that even during his time in the Sierra Nevada, all of these spectacular creatures were rendered extinct by habitat loss and state-government-sponsored cash bounties for bear killings. The last California grizzly was killed in Yosemite in 1895, and the last such bear in all of California was shot to death in Fresno County in 1922. Ironically, in spite of the state's other notable conservation efforts, California is the only state to bear the image of an extinct animal—the California golden bear or California grizzly bear—on its state seal and flag. See Allan A. Schoenherr, *A Natural History of California* (Berkeley: University of California Press, 1992), 313–405.

28. See this formulation in Bill Devall and George Sessions, *Deep Ecology: Living as If Nature Mattered* (Layton, Utah: Gibbs Smith, 1985).

29. Muir, *John of the Mountains*, 138.

30. John Muir, quoted in Richard Cartwright Austin, *Baptized into Wilderness: A Christian Perspective on John Muir* (Atlanta: John Knox, 1987), 49.

31. Muir, "Hetch Hetchy Valley," 814–15.

32. Muir, "Save the Redwoods," in *John Muir: Nature Writings*, 828. In a note about the appeal "Save the Redwoods," first published in 1920 in the *Sierra Club Bulletin*, William Cronon says that the text was "found among Muir's papers after his death" (ibid., 852). Since the wider text "Save the Redwoods" refers to a California congressional vote in 1905 to cede oversight of Yosemite and some surrounding areas to the U.S. government, it is reasonable to assume "Save the Redwoods" was written at some point in the last decade of Muir's life.

33. William Cronon, "Chronology," ibid., 840.

34. Frederick Turner, *Rediscovering America: John Muir in His Time and Ours* (New York: Viking, 1985), 70–71.

35. Evan Berry, *Devoted to Nature: The Religious Roots of American Environmentalism* (Berkeley: University of California Press, 2015), 80–81.

36. Muir, *John of the Mountains*, 138.

37. John Muir to Asa Gray, December 18, 1872, in *Life and Letters of John Muir*, 1:370.

38. Stephen Fox, *The American Conservation Movement: John Muir and His Legacy* (Madison: University of Wisconsin Press, 1981), 50.

39. Ibid., 360.

40. Michael P. Cohen, *The Pathless Way: John Muir and American Wilderness* (Madison: University of Wisconsin Press, 1984), 109.

41. Ibid., 25.

42. Ibid., 126.

43. Bron Taylor, *Dark Green Religion: Nature Spirituality and the Planetary Future* (Berkeley: University of California Press, 2015), 62 (emphasis added).

44. Ibid.

45. See Badé's reference to Muir's use of an eagle feather stylus in *John Muir: His Life and Letters and Other Writings*, 1:64; and a reference to Sequoia ink in John Muir to Jeanne Carr, ca. 1870, in *Life and Letters of John Muir*, 1:270.

46. John Muir to Jeanne Carr, ca. 1870, 1:270–73.

47. Muir, *John of the Mountains*, 138.

48. Like Taylor et al., other Muir scholars often miss this point: because everyday life is saturated with transcendence, Muir does not wonder about some other world that might exist outside of this world. Catherine Albanese, for example, sheds crucial insight on the religious influences in Muir's thought when she writes that "the Calvinist-tinged Christianity of Muir's childhood . . . did not vanish but, instead, played itself out in a different key." Albanese, *Nature Religion in America: From the Algonkian Indians to the New Age* (Chicago: University of Chicago Press, 1990) 99. But she then concludes her study by trading on the false disjunction, at least for Muir, between the "supernatural" and the "natural": "Muir had successfully taken biblical language and inverted it to proclaim the passion of attachment, not to a supernatural world but to a natural one" (ibid., 101). Muir, however, does not *reverse* two different orders of being—the supernatural and the natural—but instead proclaims the Spirit-infused beauty of *this* world in its everyday quotidian wonder. Muir does not speculate about whether there is a god or a heaven or something else outside of this world. His finely tuned biblical focus is on the full incarnation of divinity in this life. This is a focus that easily comports with the Synoptic Jesus' comment that the kingdom of God is not an otherworldly but a commonplace reality: "The kingdom of God

will not come with observable signs. . . . For you see, the kingdom of God is in your midst" (Luke 7:20–21).

49. Quoted in Fox, *American Conservation Movement*, 350.

50. Muir, *John of the Mountains*, 86.

51. See note for Jeremiah 8:22, in *The New Oxford Annotated Bible, NSRV with the Apocrypha*, 3d ed. (Oxford: Oxford University Press, 2007), 1092.

52. The same-sex character of Muir's preachment in this passage is noteworthy. Muir's call to "suck Lord Sequoia" not only violates conventional Western Christian boundaries between divinity and corporeality but also crosses lines of division between Christian heteronormativity and same-sex love and desire. Muir's ecstatic green theology celebrates both the fullness of divinity within the natural world *and* gender nonconformity in the loving practice of this reality as well. For three superb analyses of the deep affinities between Christian theology, environmental well-being, and queer identity and practice, see Daniel T. Spencer, *Gay and Gaia: Ethics, Ecology, and the Erotic* (Cleveland, Ohio: Pilgrim, 1996); Marcella Althaus-Reid, *The Queer God* (London: Routledge, 2003); and Whitney A. Bauman, *Religion and Ecology: Developing a Planetary Ethic* (New York: Columbia University Press, 2014).

53. Muir, *John of the Mountains.*

54. Augustine, *Confessions*, trans. Garry Wills (New York: Penguin, 2002), bk. 9, chap. 23.

55. Ibid., bk. 12, chap. 10.

56. John Muir to Catharine Merrill, June 1872, in *Life and Letters of John Muir*, 1:330–31.

57. Daniel Muir to John Muir, March 19, 1874, ibid., 20.

58. John Muir to Catharine Merrill, June 1872, ibid., 332–33.

59. John Muir to J. B. McChesney, January 10, 1873, ibid., 378.

### 5. ON THE WINGS OF A DOVE

1. Anne Primavesi has written widely about Earth as an incalculably valuable wonder—a "gift event" that refuses to be quantified or commodified. See her *Sacred Gaia* (London: Routledge, 2000) and *Gaia's Gift* (London: Routledge, 2003). She compares Earth as pure gift to the song of a bird: "How could I pay a bird to teach me how to sing? What cash token corresponds to its freely given song?" (*Gaia's Gift*, 111).

2. See this analysis in Gerardo Ceballos, Paul R. Ehrlich, Anthony D. Barnosky, Andrés García, Robert M. Pringle, and Todd M. Palmer, "Accelerated Modern Human-Induced Species Losses: Entering the Sixth Mass Extinction," *Science Advances* 1 (2015), http://advances.sciencemag.org/content/1/5/e1400253.full. Also see Niles Eldredge, *Life in the Balance:*

*Humanity and the Biodiversity Crisis* (Princeton, N.J.: Princeton University Press, 1998).

3. See Fred Pearce, *With Speed and Violence: Why Scientists Fear Tipping Points in Climate Change* (Boston: Beacon, 2008).

4. See Elisabeth Rosenthal and Andrew C. Revkin, "Science Panel Says Global Warming Is 'Unequivocal,'" *New York Times*, February 3, 2007, A1, A5. Also see James Gustave Speth, *Red Sky at Morning: America and the Crisis of the Global Environment* (New Haven, Conn.: Yale University Press, 2004); Elizabeth Kolbert, *Field Notes from a Catastrophe: Man, Nature, and Climate Change* (New York: Bloomsbury, 2006); and Kolbert, *The Sixth Extinction: An Unnatural History* (New York: Holt, 2014).

5. Robin Wall Kimmerer, *Braiding Sweetgrass: Indigenous Wisdom, Scientific Knowledge, and the Teaching of Plants* (Minneapolis: Milkweed Editions, 2013), 383.

6. See the excellent analysis of how the work of other natural and social scientists has been used to advance spiritual understandings of nature in Lucas F. Johnston, *Religion and Sustainability: Social Movements and the Politics of the Environment* (Sheffield, U.K.: Equinox), 78–106.

7. See the argument for the intrinsic value of all species, independent of their utility to meet human needs, in Richard B. Primack, *A Primer of Conservation Biology*, 2d ed. (Sunderland, Mass.: Sinauer Associates, 2000), 1–62.

8. James Lovelock, *The Ages of Gaia* (New York: Norton, 1988).

9. James Lovelock, *Gaia: A New Look at Life on Earth* (1979; repr., Oxford: Oxford University Press, 2000), 9.

10. Ibid., 99.

11. In addition to Lovelock, see Johnston, *Religion and Sustainability*, 78–106.

12. See Bruno Latour, *Science in Action: How to Follow Scientists and Engineers through Society* (Milton Keynes, U.K.: Open University Press, 1987).

13. See the call to preserving a just and verdant Earth as sacred work in David Suzuki with Amanda McConnell, *The Sacred Balance: Rediscovering Our Place in Nature* (Vancouver: Greystone Books, 1997).

14. Lovelock, *Gaia*, 9, 140. Shelly Rambo writes that the task of theology in the midst of suffering is to engage in a "middle discourse" between religious triumphalism, on the one hand, and the loss of faith, on the other. To witness trauma using middle discourse is to account for ongoing fragmentation and despair vis-à-vis the broken promise of redemption and renewal. Today, we are on a collision course with ourselves. Beyond the loss of species and habitats, rising sea levels are destroying the lives of millions of

human beings. Wealthy countries dump heat-trapping, ice-melting gases into the atmosphere, causing rising sea levels and massive flooding in low-lying nations such as the Maldives, Fiji, and Bangladesh and in the United States, in places such as New Orleans, Houston, the Florida Keys, and coastal New Jersey and New York City, where recent hurricanes and storms have killed and displaced thousands. In witness to this unfolding global tragedy, middle discourse theology precariously positions itself between the fractured possibility of new life and the hopelessness of despair. See Shelly Rambo, *Spirit and Trauma: A Theology of Remaining* (Louisville, Ky.: Westminster John Knox, 2010).

15. Michael Eaude, *Catalonia: A Cultural History* (Oxford: Oxford University Press, 2008), 130–31.

16. Ibid., 126–32.

17. See Ihor Kutash, "The Beauty of the Saints," Ukrainian Orthodoxy, accessed November 9, 2017, http://www.ukrainian-orthodoxy.org/saints/beauty/OnuphriusEng.htm.

18. See Gary R. Varner, *Sacred Wells: A Study in the History, Meaning, and Mythology of Holy Wells and Waters* (New York: Algora, 2009); and Walter L. Brenneman, *Crossing the Circle at the Holy Wells of Ireland* (Charlottesville: University of Virginia Press, 1995).

19. See a brief history of the Camino in the context of medieval pilgrimage in general in Maria I. Macioti, "Pilgrimages of Yesterday, Jubilees of Today," in *From Medieval Pilgrimage to Religious Tourism: The Social and Cultural Economics of Piety*, ed. William Swatos Jr. and Luigi Tomasi (Westport, Conn.: Praeger, 2002), 74–90.

20. See, for example, Kelly Lipscomb, *Adventure Guide: Spain* (Edison, N.J.: Hunter, 2005), 555.

21. Audrey and I received different *compostelas* in the Santiago cathedral at the end of our journey. We were asked in the cathedral's Oficina del Pelegrino (Pilgrim's office) what type of *compostela*, based on our spiritual intentions, we preferred. Audrey chose the nonreligious certificate of completion, composed in a flowing script with colorful images, while I chose the religious certificate, written in muted sepia tones bordered by a symmetrical clamshell design.

22. I have written about the importance of ritual, outside of traditional religious institutions, in terms of grounding exercises in everyday life, including the classroom setting, in "Experience, Purpose, Pedagogy and Theory: Ritual Activities in the Classroom," in *Teaching Ritual*, ed. Catherine Bell (New York: Oxford University Press, 2007), 73–87.

23. John Muir, *John of the Mountains: The Unpublished Journals of John Muir*, ed. Linnie Marsh Wolfe (Boston: Houghton Mifflin, 1938), 427.

24. John Brierley, *A Pilgrim's Guide to the Camino de Santiago: A Practical and Mystical Manual for the Modern Day Pilgrim (The Way of St. James / Camino Francés)* (Forres, Scotland: Findhorn, 2011), 36.

25. Jack Kerouac, *Dharma Bums* (1958; repr., New York: Penguin Books, 1971).

26. The Camino is many things to many people—history, tourism, athleticism, friendship, family, love, rural life, eating and drinking, buying and selling, and the spiritual quest—and for many, it marks a turning point in one's life. The ambiguity I experienced being a religious traveler along the route was heightened at our final destination. Audrey and I marked our completion of the Camino in the pomp and ceremony of the Compostela cathedral. After our weeklong sojourn, often being alone for hours on end in quiet contemplation and conversation, we were now surrounded by throngs of day-trippers and package vacationers, as well as other pilgrims and churchgoers, in a sanctuary that felt both private and overwhelming at the same time. Eleanor Munro summarizes my feelings at that time: "It is mostly antiquarian and touristic interest that brings crowds to Santiago today. Still, when the cathedral is full of fire, and a massed choir of hundreds begins to sing a swirling medieval hymn, and the words *pelegrino . . . pelegrino* break over one's head, it is hard not to feel that the place lives still, though not any more in our plane of time." Eleanor Munro, *On Glory Roads: A Pilgrim's Book about Pilgrimage* (New York: Thames and Hudson, 1987), 217.

27. David Haberman, *The Journey through the Twelve Forests: An Encounter with Krishna* (Oxford: Oxford University Press, 1994), 71.

28. See image of Millet's *Bird's-Nesters* at https://commons.wikimedia.org/wiki/File:Jean-François_Millet,_French_-_Bird%27s-Nesters_-_Google_Art_Project.jpg.

29. See Ben J. Novak, "The Passenger Pigeon: Ecosystem Engineer of Eastern North American Forests," Revive & Restore, December 12, 2016, http://reviverestore.org/the-passenger-pigeon-the-ecosystem-engineer-of-eastern-north-american-forests/.

30. See Barry Yeoman, "Why the Passenger Pigeon Went Extinct," *Audubon*, May–June 2014, http://www.audubon.org/magazine/may-june-2014/why-passenger-pigeon-went-extinct. Also see Jonathan Rosen, *The Life of the Skies: Birding at the End of Nature* (New York: Farrar, Straus and Giroux, 2008), 32–44.

31. In Pennsylvania, thousands of wild pigeons are killed in dozens of annual pigeon shoots, generally sponsored by local gun clubs, in a blood-sport frenzy supported by conservative lawmakers and the National Rifle Association. Might this cruelty be impacted by forging a spiritual link

between the dovey pigeon of Jesus' baptism and the rock dove (feral pigeon) victims of today's pigeon shoots? I am grateful to Audrey for suggesting this link. See John Kopp, "Pennsylvania Clings to Pigeon Shoots That Have Nearly Vanished Nationwide," *Philly Voice*, June 5, 2017, http://www.phillyvoice.com/pennsylvania-clings-to-pigeon-shoots-that-have-nearly-vanished-nationwide/.

32. W. S. Merwin, "Thanks," in *The Essential W. S. Merwin*, ed. Michael Wiegers (Port Townsend, Wash.: Copper Canyon, 2017), 205. I am grateful to Steven Hopkins for directing my attention to this radiant work of literary art.

33. See Joel Achenbach, "Humans Wiped Out Billions of These Birds," *Washington Post*, November 16, 2017, https://www.washingtonpost.com/news/speaking-of-science/wp/2017/11/16/billions-or-bust-new-genetic-clues-to-the-extinction-of-the-passenger-pigeon/?utm_term=.648c96071a41.

34. Matthew Eaton, "Enfleshing Cosmos and Earth: An Ecological Theology of Divine Incarnation" (Ph.D. diss., University of St. Michael's College of the Toronto School of Theology, 2016), 250–51.

35. The explanation of Jesus' unrecognizability in the Gospels as a result of his followers' fear or stress is conventional but unsatisfying to me. An equally plausible explanation is that Jesus' identity, while always apparent to others, appears to be veiled or shadowed at times, requiring those closest to him to actually *see* him and thereby identify him accordingly. In the work of the fin de siècle American artist Henry Ossawa Tanner, Jesus is often depicted as a dark-complected man in cloudy outdoor or dimly lit indoor settings. Tanner uses a subdued, earth-tone palette to create images of Jesus that are gentle, introspective, and strangely intimate. In Tanner, Jesus is not a transcendent God-Man but a humble, and yet remarkably present, human being. The onus is on the contemporary viewer of the painting—indeed, as it was on Jesus' disciples long ago—to realize who he is within a variety of less-than-obvious settings. No supernatural halos or artificial rays of light signal to the viewer Jesus' identity; rather, nocturnal scenes, obscure angles, and rearview perspectives characterize Tanner's compositions of Jesus' many manifestations. In *Jesus and Nicodemus* and *Nicodemus Coming to Christ*, Jesus' face is partially hooded; in *The Disciples See Christ Walking on the Water*, Jesus is seen from behind, and his figure blends in with the blue-green expanse of the sea. In *The Good Shepherd*, he is in dark silhouette underneath blackened oak trees and a pale moon. And in *He Disappeared out of Their Sight*, the postresurrection Jesus is only present in the painting in his shadow left behind after vanishing from the view of his disciples. Tanner's biblical paintings are not displays of Jesus' God-like power but rather subtle invita-

tions to enter into the reality of Christ's existence through patience, humility, and keen attention to detail. Tanner's message, I think, is that Jesus is here, but his presence is not as obvious as conventional piety might assume. Rather, the discovery of who Jesus is is hard won and richly rewarding for the viewer, the follower, who enters into an attentive relationship with his person. See Anna O. Marley, ed., *Henry Ossawa Tanner: Modern Spirit* (Berkeley: University of California Press, 2012) for beautiful color plates and thoughtful analyses of Tanner's extraordinary aesthetic.

36. The Swiss reformer John Calvin's notion of common grace, for example, entails the presumption that within every human being there exists, as he alternately puts it, "by natural instinct, an awareness of divinity," "a sense of deity inscribed in the hearts of all," or "a seed of religion in all men." John Calvin, *Institutes of the Christian Religion*, ed. John T. McNeill, trans. Ford Lewis Battles, 2 vols. (Philadelphia: Westminster, 1977), 1:39–63.

37. See the full text at http://www.vatican.va/archive/hist_councils/ii_vatican_council/documents/vat-ii_const_19651118_dei-verbum_en.html.

38. Kolbert, *Sixth Extinction.*

39. Chelsey Harvey, "Greenland Lost a Staggering 1 Trillion Tons of Ice in Just Four Years," *Washington Post*, July 19, 2016.

40. See this extracanonical exhortation to Christian asceticism from late antiquity, urging its readers to cleanse themselves of worldly pollution by overcoming fleshly temptations: "Blessed are those who have not polluted their flesh by craving for this world, but are dead to the world that they may live for God! "Pseudo-Titus," in *Lost Christianities: Books That Did Not Make It into the New Testament*, by Bart D. Ehrman (Oxford: Oxford University Press, 2003), 239.

41. Edward Wong, "Trump Has Called Climate Change a Chinese Hoax: Beijing Says It Is Anything But," *New York Times*, November 18, 2016.

42. Gregory A. Smith and Jessica Martínez, "How the Faithful Voted: A Preliminary 2016 Analysis," Pew Research Center, November 9, 2016, http://www.pewresearch.org/fact-tank/2016/11/09/how-the-faithful-voted-a-preliminary-2016-analysis/.

43. William James, *The Varieties of Religious Experience: A Study in Human Nature* (1901; repr., New York: Seven Treasures, 2009), 51–95.

44. Gayatri Chakravorty Spivak, *A Critique of Postcolonial Reason* (Cambridge, Mass.: Harvard University Press, 1999), 355n59.

45. John Milton, *Paradise Lost*, bk. 4, v. 678.

46. Elizabeth Barrett Browning, *Aurora Leigh*, bk. 7, v. 86.

47. Gerard Manley Hopkins, "God's Grandeur," in *Burning Bright: An Anthology of Sacred Poetry*, ed. Patricia Hampl (New York: Ballantine, 1995), 32.

48. As a final coda, I am reminded here of the animocentric sensibility in the book of Job, reminiscent of Muir's Jobian-ursine Christianimism, where God says, "But ask the animals and they will teach you; the birds of the air and they will tell you; speak to the Earth and it will teach you; and the fish of the sea will declare to you" (Job 12:7–8).

# INDEX

**groundworks|**

**ECOLOGICAL ISSUES IN PHILOSOPHY AND THEOLOGY**

**Forrest Clingerman and Brian Treanor, series editors**

***Interpreting Nature: The Emerging Field of Environmental Hermeneutics***
Forrest Clingerman, Brian Treanor, Martin Drenthen, and David Utsler, eds.

***The Noetics of Nature: Environmental Philosophy and the Holy Beauty of the Visible***
Bruce V. Foltz

***Environmental Aesthetics: Crossing Divides and Breaking Ground***
Martin Drenthen and Jozef Keulartz, eds.

***The Logos of the Living World: Merleau-Ponty, Animals, and Language***
Louise Westling

***Being-in-Creation: Human Responsibility in an Endangered World***
Brian Treanor, Bruce Ellis Benson, and Norman Wirzba, eds.

***Wilderness in America: Philosophical Writings.* Edited by David W. Rodick**
Henry Bugbee

***Eco-Deconstruction: Derrida and Environmental Philosophy***
Matthias Fritsch, Philippe Lynes, and David Wood, eds.

***Animality: A Theological Reconsideration***
Eric Daryl Meyer

***When God Was a Bird: Christianity, Animism, and the Re-Enchantment of the World***
Mark I. Wallace

CPSIA information can be obtained
at www.ICGtesting.com
Printed in the USA
JSHW021104100821
17743JS00002B/182

9 780823 281312